DATE DUE

THE GENIE IN THE MACHINE

THE GENIE
IN THE MACHINE

*How Computer-Automated Inventing
Is Revolutionizing Law and Business*

Robert Plotkin

STANFORD LAW BOOKS
An Imprint of Stanford University Press
Stanford, California

Stanford University Press
Stanford, California

Printed in the United States of America on acid-free, archival-quality paper

Library of Congress Cataloging-in-Publication Data

Plotkin, Robert, 1971-
  The genie in the machine : how computer-automated inventing is revolutionizing law
and business / Robert Plotkin.
     p. cm. — (Stanford law books)
  Includes bibliographical references and index.
  ISBN 978-0-8047-5699-0 (cloth : alk. paper)
  1. Patent laws and legislation—United States. 2. Inventions—United States. 3. Computer programs—Patents. 4. Intellectual property—United States. I. Title.
  KF3131.P58 2009
  346.7304'86—dc22
                        2008040840

Typeset at Stanford University Press in 10/14 Minion

To my parents, for making all of my wishes come true

Contents

THE GENIE IN THE MACHINE

Introduction

The Coming Artificial Invention Age

THERE IS AN UNSEEN COMPUTER REVOLUTION under way: the revolution in computer-automated inventing. Computers are now designing products in ways that previously required human ingenuity, thereby ushering in a new era that I refer to as the Artificial Invention Age.[1]

This isn't Steven Spielberg's next science fiction movie. Artificial invention technology[2] is already here, and we're already buying and using its creations. For example, Stephen Thaler of Imagination Engines used computer software called the Creativity Machine, which is modeled after the creative processes of the human brain, to invent the crossed-bristle configuration of the Oral-B CrossAction toothbrush. One user of the Creativity Machine has described it as "Thomas Edison in a box."[3] Gregory Hornby of the NASA Ames Research Center used "evolutionary" software to dream up a tiny antenna—a weird little object looking for all the world like an unwound paper clip—that is now on a space mission (see Figure 1).[4] He admits that no human engineer would have thought of an antenna that looked so crazy, yet the antenna works better than previous human designs.[5] John Koza of Genetic Programming used "genetic programming" software to create a new controller, a kind of device found in everything from thermostats to automobile cruise control systems. The proof of the controller's novelty is in the pudding—or perhaps I should say in the patent that the U.S. Patent Office granted not only on the controller itself but also on the computer-automated method that Dr. Koza used to invent it.[6]

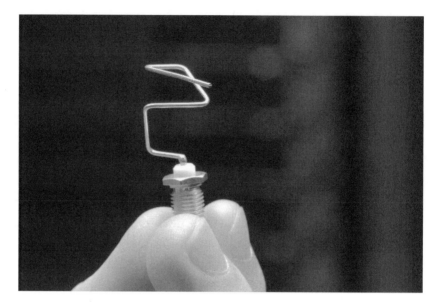

FIGURE 1 Antenna in space. Courtesy NASA.

Genies in a New Guise

Software, not a human inventor, devised the designs for each of these products.[7] The latest generation of artificial invention software therefore differs qualitatively from all previous tools inventors have used to assist them in inventing, from the very first stone with which a human sharpened a branch into a spear, to the computer aided design (CAD) software used by today's automotive engineers to construct 3D models of tomorrow's car engines. Even the most advanced technology of the Industrial Age could not conceive the shape of a new airplane wing, evaluate the efficiency of airflow around that wing, and then modify the wing's design to make it even more aerodynamic. Such acts could only be performed by a human mind.

Not any more. Dr. Thaler did not tell the Creativity Machine to use a crossed-bristle configuration for the CrossAction toothbrush. Instead he fed existing toothbrush designs into the Creativity Machine and then gave the Creativity Machine objective data about how effectively each of those toothbrushes cleaned teeth. Solely from this information, the Creativity Machine discovered what makes one toothbrush better than another at brushing teeth, and produced the crossed-bristle design on the basis of that discovery.[8] Nor did Dr. Koza tell his genetic programming software which components to use in the controller he patented. Instead, he merely told the software which *criteria* he needed a controller to satisfy, and in response the software *automatically* devised a controller that satisfied those criteria.

This will be the role of human inventors in the Artificial Invention Age: to formulate high-level descriptions of the problem to be solved, not to work out the details of the solution. Filling in those minutiae is precisely the task at which artificial invention technology excels.[9] In this sense, a computer running artificial invention software is like a genie, and the problem description that the human inventor supplies to the software—such as "generate an antenna that weighs less than a pound and can transmit and receive FM radio signals"[10]—is like a wish.[11] Once given this problem description (wish), the artificial invention software (genie) produces a design for a concrete product—a toothbrush, an antenna, a controller—that solves the stated problem. Such a design is the inventor's wish come true. This *fundamental structure of a wish*, shown in Figure 2, is evidence of Arthur C. Clarke's Third Law: "Any sufficiently advanced technology is indistinguishable from magic."

Human and Computer: A New Partnership

Artificial invention technology will reduce the amount of time and money required to produce new inventions, but that will not be its most profound effect. As the photo of the NASA antenna makes clear, artificial inventions generated by software often appear bizarre to human eyes, even those of technical experts. Today's software can even produce inventions that human inventors previously declared unattainable. For example, John Koza used his genetic

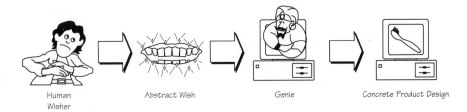

| Human Wisher | Abstract Wish | Genie | Concrete Product Design |

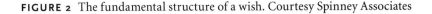

FIGURE 2 The fundamental structure of a wish. Courtesy Spinney Associates

programming software to produce a circuit consisting only of resistors and capacitors with a gain (amplification factor) of greater than two, even though skilled electrical engineers in the field had claimed that such a circuit could not be created.[12] One reason computer-generated designs often confound human experts in this way is that artificial invention software lacks the biases and blind spots that can keep human inventors from considering possibilities that are fruitful but violate conventional design wisdom.[13] Artificial invention technology therefore not only allows inventors to focus on high-level design rather than low-level details; it also enables them to produce inventions that they could not have created *at all* without such technology.

Note that I say invention automation technology enables *human inventors* to invent more effectively, *not* that the technology itself has become an inventor. Although it certainly is true that artificial invention technology can automate significant *parts* of the inventive process—such as selecting and combining components into a final design—this does not mean such technology will make human inventors obsolete in the Artificial Invention Age. Rather, human inventors will remain in the driver's seat for the foreseeable future, formulating wishes and using them to steer the technology toward its final destination. Today's artificial invention technology can invent nothing of its own volition, just as a genie can only sit idly by without an Aladdin to make a wish.

The fact that artificial invention technology automates the inventive process only partially does not diminish its significance. First, as we've already seen, even such partial automation automates the kind of physical design processes that, until now, were the exclusive province of human creativity. Second, every kind of machine-automated process we've ever encountered is partial in the sense that it requires some human involvement to guide it. Yet the incompleteness of such automation does not make it unworthy of attention. Take the automobile, whose very name means "automated motion machine." No one today would deny that the automobile did in fact automate transportation in a noteworthy way, even though if you sit down in a car it will take you nowhere on its own volition. You, the human operator, must still control its every turn. If you get stopped for speeding, your plea that the car exceeded the speed limit "automatically" will fall on deaf ears.

This is just one example of a more general phenomenon: that humans and machines always interact with each other as a system, even when those machines automate a task previously performed by humans. Inventors and the tools they use are no exception. Long before there were computers, inventors used physical tools, ranging from hammers to line levels to scales, to help them build, test, and measure their inventive works-in-progress. Inventors are also expert at leveraging conceptual tools (such as the laws of physics expressed in mathematical language) to help them design new machines, predict their performance, and understand their reasons for failure. It is hard to imagine inventors engaging in any but the simplest inventive activity without using tools.

Inventors are masters at using tools, but also at inventing new tools to boost their own inventive abilities. Early humans learned to use one rock to sharpen another, and then to use that newly sharpened rock to carve a boat from a fallen tree. Savvy inventors will use artificial invention technology in the same way, to boost their inventive abilities to previously undreamt heights. Focusing exclusively on the possibility that computers will *replace* human inventors overlooks the more complex, probable, and promising prospect that inventors will leverage computers to *augment* their inventive skills and thereby surpass the inherent limitations of their own minds.

Therefore, although today's artificial invention technology is the next step in the age-old evolution of tools used by inventors, it is a step with profound consequences for how we invent because it will produce a qualitative shift in the division of labor between human inventors and their tools, and a corresponding increase in the power of the human-computer inventive system. Human inventors, once responsible for designing every detail of their inventions, will be freed (and highly valued for their ability) to identify the problems they are trying to solve, and to pose those problems to artificial invention software in a language the software can understand. The software will then do the rest.

Genies v1.0 and Beyond

Today's artificial invention technology is a logical extension of the technology that computer programmers have been using for decades to create software. Unlike previous generations of innovators, who forged their inventions by molding, cutting, and extruding raw materials to match their designs, computer programmers bend computers to their will by writing computer programs consisting of instructions. Then they provide those instructions to a computer, which automatically transforms the instructions into working software in the computer's memory or on the CD that you buy in a store. In this sense, computers have *always* been like genies, transforming the wishes of programmers into reality.

Getting a genie to do your bidding isn't quite as simple, however, as just putting your wish into words. Anyone who's read a story about genies or wishing (think of the myth of King Midas) knows that wishing can be a tricky—and even dangerous—business.[14] One slip of the tongue and you may find yourself turned into a toad rather than bathed in riches. Successful wishing requires you to phrase your wish using just the right magic words. Computer programming works the same way, requiring mastery of programming languages whose rules must be followed with extreme precision. Every time your computer crashes you can rest assured that some software on your computer was written by a programmer who didn't follow the wishing rules to a T. Similarly, although inventors in the Artificial Invention Age will not need to muck around with the low-level details of their inventions, they will still need expertise of another kind to write instructions that can successfully command a computer to produce working inventions.

We've all witnessed firsthand the benefits of this early version of computers as artificial genies in contexts extending far beyond our desktop computers. Manufacturers of everything from televisions to microwave ovens to radar systems have been replacing old mechanical and electrical components with software for decades, not just because software is cheaper to manufacture and distribute but because software is faster, easier, and less expensive to design. Why pay an electrical engineer to design a complex circuit for calculating a mathematical formula when a computer programmer can do the same job by writing one line of computer code, a one-sentence wish?

The same basic technique that automatically transforms the instructions written by programmers into working software is also used to create the microprocessors of modern computers themselves. No human engineer or team of engineers could hope to design manually the billions of transistors in a modern microprocessor. Instead, electrical engineers design a new processor by writing instructions in a "hardware description language" (HDL) explaining what they want the processor to *do* and then give the description to software that transforms the description into a schematic for a processor.[15] Tomorrow's engineers will use similar techniques to design the next generation of nanotechnology precisely because of the complexity of the technology and the benefits of computer-automated inventing.

Inventing by wishing is here to stay. Combine artificial invention technology's ability to automatically create machine designs with continuing improvements in low-cost automated manufacturing and networked collaboration, and you'll see we stand poised for an exponential expansion of the ability to satisfy our material needs.

A Roadblock: Patent Law

Yet there's a roadblock standing in the way of this bright future: patent law. Today, patent law determines *who owns inventions*. Tomorrow, patents on artificial invention technology will determine *who owns the right to invent*. Patent law will therefore confer awesome power on those who take advantage of it—too much power, if we don't begin to update patent law now.

Just imagine if Henry Ford had been given absolute control over who could and could not use an assembly line to speed production and lower costs, or if telegraph inventor Samuel Morse had obtained the power to block anyone else from using not just the telegraph but *any* machine capable of transmitting messages over long distances using electricity. (In fact, Morse attempted to obtain just such a broad patent but was shot down by the U.S. Supreme Court in 1854.[16]) Granting powerful patents to inventors of early advances in artificial invention technology could bestow on such patent owners the ability to block competitors, and the public more generally, from using such technology to produce their own inventions. Such patents, if granted improvidently, could concentrate in the hands of a few the ability to use what may turn out to be the most effective techniques for inventing in the 21st century. In this scenario, patent law would stifle innovation instead of promoting it.

We should not, however, rule out patents on artificial invention technology or the inventions it produces simply because such patents could be abused. Other new technologies, from traditional software to gene sequencing, have required us to rethink how we should apply patent law to them to ensure that the resulting patent rights are neither too weak nor too strong, since erring in either direction can impede technological progress. The advent of the Artificial Invention Age brings us to another such inflection point. If a computer running powerful artificial invention software is like a genie, and if the instructions that a human gives to such a computer are like a wish, then we must squarely face the question, Should artificial genies, or the wishes they grant, or the inventions they produce, be patentable?

My answer to all three parts of this question is a qualified yes, because all such patents will play an important role in promoting innovation in the Artificial Invention Age, so long as such patents are examined particularly carefully to ensure that only the right ones are granted. Recall the Creativity Machine, which Stephen Thaler used to design the CrossAction toothbrush. Dr. Thaler obtained a patent on the Creativity Machine, in which he described how the Creativity Machine works. Now anyone with the right technical background can read the patent and—once the patent expires—build and use her own Creativity Machine *for free*. Dr. Thaler would have had little reason to publish such a description, given that his competitors could have used it to compete against him, were it not for the patent rights that he received in exchange. In this way, the patent system performed its intended role of expanding the public's knowledge about a significant technological advance.

As people such as Dr. Thaler develop even more powerful artificial genies, the lure of patent protection for those genies will encourage them both to continue innovating and to describe their inventions in patents, which are published for the world to read. If we deny patent protection to artificial invention technology, or confer protection that is too weak, the Stephen Thalers of the world may forgo inventing the next Creativity Machine, or keep it a secret, making money by charging companies for the right to use it in private to generate new inventions. In this scenario, too, the public loses because the only players who will benefit from artificial invention technology are those who can afford to pay for the right to use it.

Standing at the outset of the Artificial Invention Age, therefore, we must ensure that patent law strikes the right *balance* when allocating ownership rights in artificial invention technology and the inventions it produces. Patent rights, like Goldilocks's porridge, must be neither too strong, nor too weak, but just right. If we fail to strike the right balance, the cost to our economy, to the growth of technology, and even to the quality of our lives will be incalculable.

We have every reason to believe, however, that patent law as it exists today will strike the *wrong* balance if we apply it to artificial invention technology. Those who crafted the patent system over the course of centuries never dreamed that machines, rather than humans, would one day design inventions. Like a car driven at twice its rated speed limit, patent law will break down if it is applied to artificial invention technology.

We can already see the cracks in the dam as patent law rattles even in the face of old-fashioned software. Part of the problem is the sheer number of patent applications for software. About 15 percent of *all* patents granted in the United States are now for software, with the number of such applications increasing eighteenfold over 20 years.[17] An inventor who files a patent application often waits for years until a patent examiner even reads the application for the first time. Government patent offices simply can't process this flood of patent applications effectively, and artificial invention technology will only open the floodgates wider.

One reason for this increase in software patent applications is that companies are becoming increasingly savvy about using patents as sword and shield in battle against their competitors. Large companies routinely obtain software patents as trading chips with other companies, or to block competitors; small companies patent their software to secure their investment and protect themselves against copying by bigger companies. More controversially, some small companies don't invent at all; they simply round up obscure software patents so they can collect royalties, licensing fees, and judgments from other companies using the technology.[18] Both Microsoft and Research in Motion (manufacturer of the Blackberry) learned this lesson the hard way when they were sued for patent infringement by small companies and each paid over *half a billion dollars* as a result.[19] There is every reason to think that companies will apply the same strategies to artificial invention technology and the inventions it produces.

The response of the courts to the ever-widening vortex of litigation generally has been to expand the range of software that can be patented, without acknowledging the role of automation in the process of inventing software, thereby escalating the problem instead of bringing it under control. Meanwhile, patent offices worldwide have done little to reform their software patent practices.

There is, however, a deeper problem with software patents. Patent law—which was designed to apply to concrete inventions such as the cotton gin and light bulb, whose physical structure we could understand and describe clearly and directly—still has not figured out how to grapple with something as seemingly ephemeral as software.

This problem will only intensify once patent law wrestles with artificial invention technology. It will be like trying to define legal rules for patenting the genie and the wish, rather than the lamp. Should Stephen Thaler's patent on the Creativity Machine enable him to prohibit anyone else from using the Creativity Machine to invent anything? Should the patent give Dr. Thaler ownership of *all* of the toothbrushes that the Creativity Machine invents, or *could* invent—even if there are thousands of them? Dr. Thaler told me that although he could flood the Patent Office with such patent applications, so far he has chosen not to do so. What happens when he and others in the field of artificial invention decide to stop playing Mr. Nice Guy?

Several factors are converging to produce a perfect storm of legal conflict in this field. First, it is only in the last few years that artificial invention technology has become capable of reliably producing useful real-world inventions, rather than academic curiosities. Second, private for-profit companies based on artificial invention technology are now springing up around the world. Third, massive government and private funding of nanotechnology and biotechnology will drive further development and use of artificial invention technology because human designers cannot handle the complexity involved in manually designing the next generation of inventions in such fields.

Companies have already begun to obtain patents on artificial invention software and on inventions generated using it. The history of patents in other technological fields tells us that it is only a matter of time before disputes over ownership of such patents reach boardrooms and courtrooms in large numbers. If we begin *now* to reshape patent law to fit the new inventive paradigm made possible by artificial invention technology, the winners and losers will be sorted out in a way that promotes innovation and helps to usher in the most promising incarnation of the Artificial Invention Age.

Adapting to Invention Automation

The future, however, won't wait for patent law. Even if patent law remains unchanged in the Artificial Invention Age, everyone who comes into contact with artificial invention technology will need to learn how to adapt to it—and soon. The biggest advances are very nearly upon us, with effects potentially as profound as those of the Industrial Age. Individual programmers, scientists, and engineers who don't currently use the latest artificial invention technology (and this is the vast majority) will need to become adept at using it if they want to avoid being replaced by it. High-tech companies need to begin now to prepare for the effects of artificial invention technology on their businesses. Such companies not only need to get better acquainted with how such technology works and what kinds of teams must be assembled to exploit it;

they have to revisit their entire patent strategy, since the typical, knee-jerk approach of patenting everything that moves may or may not work in the Artificial Invention Age.

Patent lawyers like me will also need to get their minds around artificial invention technology so that they can best advise their clients. In particular, innovative companies will need to adopt nuanced legal strategies that take the operation and effects of artificial invention technology into account, whether they are obtaining patents on artificial invention technology or attempting to defend themselves against such patents.

Artificial invention technology even promises to enable consumers to become active "prosumers": producers *and* consumers of technology. Forward-looking companies, in a break from the traditional "take what we've got to sell you and be happy with it" model of doing business, are already engaging and even *collaborating* with their customers to create their next generation of products. Such developments, combined with artificial invention software and advances in low-cost "desktop manufacturing" technology, promise to bring true inventive power to the masses.

Such exhilarating possibilities, however, will only come to pass if patent law does not stand in the way. If we do nothing, control over artificial invention will fall to those players who are savvy enough to game the existing system to their private benefit. To avoid this, patent law must be reformed; it *can* be reformed, and now you will learn how.

I AUTOMATING INVENTION

1   A Prehistory of Genies

MY FIRST ENCOUNTER WITH A GENIE came when I was in fifth grade at P.S. 206 in Brooklyn, New York, in 1981, when our school obtained its first personal computers: two Radio Shack TRS-80s to be shared among the hundreds of students in the school.[1] Even by the now-primitive standards of the day, the TRS-80s were known derisively as "TRaSh 80s" for their limited memory and blocky black-and-white graphics. Our first computer lesson involved each of us walking to the back of the classroom to sit in front of one of the TRS-80s that had been wheeled there on carts as they made the rounds through the school. When my turn came, our teacher, Mrs. Kraushar, told me to type in this two-line program:

```
10 PRINT "MY NAME IS ROBERT "
20 GOTO 10
```

Next, I was instructed to type this command:

```
RUN
```

The "RUN" command in BASIC is like a textual "Go!" button, which instructs the computer to run or execute the program stored in its memory—in this case, the two-line program I had just typed in. (On a modern computer with a graphical user interface, you typically run a program by double-clicking on its icon rather than by typing a RUN command.) Typing

RUN and hitting the <ENTER> key caused the following to appear on the screen:

```
MY NAME IS ROBERT MY NAME IS ROBERT MY
 NAME IS ROBERT MY NAME IS ROBERT MY N
AME IS ROBERT MY NAME IS ROBERT MY NAM
E IS ROBERT MY NAME IS ROBERT MY NAME
IS ROBERT MY NAME IS ROBERT MY NAME IS
 ROBERT MY NAME IS ROBERT MY NAME IS R
OBERT MY NAME IS ROBERT MY NAME IS ROB
ERT MY NAME IS ROBERT MY NAME IS ROBER
T MY NAME IS ROBERT MY NAME IS ROBERT
MY NAME IS ROBERT MY NAME IS ROBERT MY
 NAME IS ROBERT MY NAME IS ROBERT MY N
AME IS ROBERT MY NAME IS ROBERT MY NAM
```

The cascade of words flowed across the screen endlessly, creating a hypnotic dance until I was eventually instructed to hit the <BREAK> key to bring the program to a halt. I was transfixed, and I like to think not only because (to paraphrase Dale Carnegie) the sweetest sight to anyone's eyes is his own name. Instead, I was struck by the fact that I had made this mysterious machine do my bidding just by typing in two simple instructions and RUN. My first artificial wish had been granted, and I was hooked.

Looking back on it, I'm glad that my first interaction with a computer, unlike that of the vast majority of children and adults today, was not to play a videogame or even to *use* software of any kind, but rather to *create* software and see the results, even if the program itself had been written by someone else for me. This imprinted on me the idea (even if I didn't recognize it at the time) that a computer is a machine whose purpose is to transform a set of written instructions into reality. Over time I realized that I could use computers to extend the reach of my will not by becoming physically stronger or more agile but by learning how to instruct computers in the languages they spoke.

I was particularly receptive to the means by which computers conferred power—written instructions—for two reasons. First, I was never adept at physical tinkering. I had no natural talent with Lincoln Logs or Tinker Toys. To this day, I shy away from attempting to fix anything in my house for fear of bringing the entire building crashing down around me. My partner loves to say to me, "You mean to tell me that you have a degree from MIT and you're a patent lawyer but you can't fix an *X*?" where *X* can be anything from a leaky faucet to a squeaky stair. "I'm software, not hardware," I reply. If I could fix a faucet by debugging its code, I'd be the next Bob Vila.

Second, my main hobby as a teenager other than computer programming was creative writing. Although the process of writing computer programs and writing fiction felt similar to me, I was keenly aware of a critical difference between the two. I could make my latest program come to life merely by typing RUN; not so with my latest manuscript. For that, I had to gather my friends together to volunteer as actors, which meant coordinating everyone's inconsistent schedules and finding that people wanted to improvise rather than follow my script as precisely and obediently as my computer would do in carrying out my programs. How I longed for a manuscript genie, to which I could simply hand my manuscript and issue the command ENACT in order to see my latest sketch performed before my eyes. (My adolescent mind could not conceive that someone else's improvisation might improve on my already-perfect script.)

As much as we mocked those old TRS-80s, they enabled someone like me—a bit of a control freak with a natural aptitude for writing and thinking in terms of logical instructions rather than nuts and bolts—to make a machine do my bidding, even if I had no idea *how* the machine did so. In those years, I wrote many programs for my own use and enjoyment to do things such as keep track of my comic book collection, organize my hard disk drive, and play videogames. Although my youthful efforts were a far cry from using computers to design antennas for space missions, my early experiences with computers as tools for boosting creative power have driven my fascination with computers ever since.

Those early experiences may also have primed me to be attuned to the power of computers as invention automation machines when I became a practicing patent lawyer. Yet the role of computers in inventing is often overlooked, or at least underemphasized in relation to its significance. We often use the term *computer revolution* to refer to the exponential increases in calculation and communication speed, storage capacity, and miniaturization effected by computer technology in just a half-century. Many excellent histories of computers situate the historical development of computers in the stream of human efforts to automate calculation, ameliorate human drudgery, and create general-purpose programmable machines. Journalistic accounts of 20th century advances in computing tend to focus on the intellect, personalities, and struggles of computing pioneers and the technologies and companies they created.

Although all of these threads are worthy of the attention they receive, no single one of them fully captures how computers have been used to assist in—and eventually to automate—significant portions of the process of invention. Yet we must place computer technology in its historical context as an invention-assisting tool if we are to evaluate the impact of today's invention automation technology on the process of invention and on patent law—the law of invention. Let me, then, take you on a tour through several interlocking histories of computers that take us as far back as the origins of life on Earth, pass through my grade school TRS-80 experience, and race onward to today's latest and greatest invention automation technology.

Software: From Hard to Soft

Software today is largely invisible to us. We download it from the Internet in the form of unseen electrical signals that course directly through our telephone lines—or even through the air—into computer memory, where those signals again take no physical form that we can observe directly. Even when we purchase software on a CD or other physical medium, we cannot discern the software from the medium itself; just try visually distinguishing one CD from another on the basis of the software they contain. To many, this is precisely what makes software "soft": its apparent intangibility in contrast to the physical hardware that the software controls. John Perry Barlow, a former Grateful Dead lyricist and member of the digerati, put this point most eloquently in his 1996 "Declaration of the Independence of Cyberspace":

> There is no matter [in cyberspace];[2] all the goods of the Information Age [including software] . . . will exist either as pure thought or something very much like thought: voltage conditions darting around the Net at the speed of light, in conditions that one might behold in effect, as glowing pixels or transmitted sounds, but never touch. . . .[3]

Although there is some truth and value to framing the distinction between software and hardware as the difference between information and matter—or, as Nicholas Negroponte of the MIT Media Laboratory famously coined it, the distinction between bits and atoms[4]—this difference obscures an equally important, but often overlooked, distinction between software and hardware.

Software has not always been stored as energy; in fact, it was embodied exclusively in matter until relatively recently in history. Therefore, software's softness in comparison to hardware must derive from some other, more fundamental, property.

The Ropebot

The great Greek engineer Hero invented what may have been the first programmable machine at around the time Christ was born: a device that could roll across the floor toward an audience and then pause, after which the top half of the machine would open and display a short automated puppet show featuring Dionysus, the Greek god of wine. On finishing the show, the device would turn around and roll back off stage.[5] Such an automated puppet theater would have been a marvel for its time just by virtue of being able to pull off such a complex stunt without human intervention. Hero, however, outdid himself by making the motions of the device *programmable*, meaning that a human operator could change the course the machine took to and from center stage merely by rewiring a portion of the machine before the performance, but without needing to redesign or rebuild the entire machine from scratch.

It isn't quite accurate to say that one "rewired" the Ropebot to reroute its path; it relied on an ingenious use of string rather than wire. The entire device was powered by a weight that lowered gradually under the force of gravity. The weight was connected to a length of twine which was wrapped around the contraption's front axle. As the weight lowered, it would unwind the twine from the axle, causing the axle to turn and thereby to rotate the device's two front wheels. The device would roll forward.

The first mechanism that made the Ropebot programmable was a peg extending from the center of the front axle. To make the Ropebot move forward a bit and then roll backward, you would wrap the string around the axle a few times in one direction, then wrap the string around the peg to enable you to continue wrapping more of the string around the axle in the *opposite* direction. Then, when you released the weight, it would pull the string off of the axle first in one direction and then the other, causing the machine to roll forward and then backward. The more times you wound the string in the first direction, the longer it would roll forward, and vice versa.

Although the Ropebot was not programmed using instructions written in text, you can *think* of a particular pattern of windings around the Ropebot's axle as representing instructions in a written programming language.

Wrapping the twine around the axle in one direction 10 times and then in the other direction five times can be expressed by a written program: "Forward (10), Backward (5)."[6] The Ropebot even included additional mechanisms enabling it to be programmed to pause, turn left, and turn right at specified points during its trip. As a result, any particular pattern of windings of rope around the Ropebot's axle can be described by a program consisting of a particular sequence of "Forward," "Backward," "Pause," "Turn left," and "Turn right" instructions.

This is the very definition of a *program*: a sequence of instructions that control a machine.[7] A "programmable machine," therefore, is one whose actions can be controlled by a program. For a program to control a machine, however, the program must take some physical form that is capable of causing the machine to perform the actions ("Turn left," "Turn right") that the program specifies. In the case of the Ropebot, the program is embodied in the form of rope wound around the axle in a particular pattern. The term *program*, therefore, is somewhat ambiguous, because it can refer either to written instructions ("Turn left") or to the (often quite different) physical form in which those instructions are embodied—such as wound rope—to control a particular machine.[8]

To avoid such ambiguities and the confusion that can arise from them, I will use the term *program* throughout this book to refer to human-understandable instructions—such as "Forward" or "Turn right," whether spoken, written, or merely conceived—and *software* to refer to instructions stored in a physical form that can control a machine, such as rope wrapped around the Ropebot's axle in a particular pattern.[9] Computer scientists sometimes call software "computer-executable instructions" to emphasize the essential feature of the latter meaning.[10]

Programming Patterns

The rope software stored in a Ropebot was inextricably intertwined with the physical mechanism of the Ropebot itself; remove the rope from the Ropebot and the software would dissipate. Contrast this with a computer today, into which you can insert a CD, run software from the CD, and then remove the CD with the software still intact on it. French inventor Joseph Marie Jacquard took a step in this direction when, in 1801, he invented what has been known ever since as the Jacquard Loom.[11] To weave a particular pattern in cloth using the Jacquard Loom, you would first punch holes into specially designed cards in an arrangement corresponding to the pattern you wanted to weave. Then you would feed the cards into the loom, which would weave the specified pattern automatically. Changing the pattern on the cards would cause the loom to weave another design.[12] A particular stack of punched cards with a specific pattern of holes on it therefore was a kind of software for the loom. In addition to being a technical marvel, the Jacquard Loom was an unbridled commercial success; in France alone 11,000 Jacquard Looms were in use by 1812.[13]

By creating a system in which the software was separated physically onto a set of cards, Jacquard introduced a degree of indirection into the process of programming his loom that the Ropebot lacked. Recall that to program the Ropebot you needed to manually wind the ropes around the Ropebot's pegs. Not so with a Jacquard Loom; to program it, you needed merely to punch holes in a set of cards and then feed those cards into the loom. In fact, you could create a set of punched cards and then hand those cards off to someone else to feed into the loom.[14] You could be a successful Jacquard Loom programmer without ever touching the loom itself. Nor did you need to be familiar with the physical mechanism by which the loom read your punched instructions from the cards to weave the patterns you specified into cloth. Instead, you needed to understand only how to write programs using the loom's "programming language"—in other words, how to punch patterns of holes into cards that would achieve the cloth pattern you desired. In this sense, Jacquard's introduction of programming by punchcard separated or *decoupled* the task of writing programs from the process of physically programming the machine itself.

Tabulation Automation

Punched cards continued to play a significant role in programmable machines well into the 20th century. Machines based on punched cards received a boost in popularity when the U.S. Census Office selected such a machine, invented by Herman Hollerith (whose company later merged with two others to form IBM) to tabulate the results of the 1890 U.S. census.[15] The Census Office had held a public contest for a more efficient way to count and interpret census results after it took seven years to tally the results of the 1880 census, leading the Office to realize that it would be impossible to count the results of the 1890 census by the turn of the century without a faster system. Hollerith won the contract by demonstrating that his machine could count results more than twice as quickly as either of its competitors.[16] His machines encoded census data about a particular individual—such as his or her race, sex, and nationality—on a punched card.[17]

The Hollerith tabulator's premier mission was a resounding success. Census workers in the field completed paper forms for about 13 million households, which were then translated into punched cards back at the Census Office.[18] Although census workers punched holes in the cards with the aid of a hole-punching machine, the act of punching holes still was largely a manual task. The benefits of the Hollerith tabulator really kicked in only when it was used to *read* data from the cards and count the results. To tabulate the results stored on cards, operators would insert the cards into the tabulator, which—much like the Jacquard Loom—would press pins against each card. Each pin that passed through a hole in a card would connect to a cup filled with mercury, thereby completing an electrical circuit and increasing a counter associated with the question corresponding to that particular hole position. Because each card contained the answers to many questions, the tabulator could count the answers to all such questions simultaneously and accurately—a significant improvement over the previous manual method of tabulation.[19] According to *Electrical Engineer* in 1891, the Hollerith tabulator "works unerringly as the mills of the gods, but beats them hollow as to speed."[20] Having proved its ability in the census, the Hollerith tabulator, and subsequent improvements to it, continued to be used by private companies such as railroads, insurance companies, and utilities to satisfy their own data processing needs for decades.[21]

Although the punched cards in such early machines were used solely to store data, such as the answers to census questions, eventually machines were developed for which punched cards could store *programs*—sequences of instructions to control the actions of the machines. For example, the Harvard Mark I computer, completed in 1944 by Howard Aiken, could process programs and data provided to it on punched paper tape, punched cards, or dial switches set manually.[22] Programmers would write their programs on paper, translate those instructions into patterns of holes on punched cards, and then feed the cards into the machine, which would read the instructions from the cards and execute them.

Plugging Programs
Although Hollerith tabulators could process census data at blinding speed, modifying the tabulator to process *different* kinds of data—such as railroad ticket sales—required tediously rewiring the tabulator itself. To address this

limitation, the Hollerith tabulator was modified in 1902 to include a plug-board similar to the kind found in old-fashioned telephone switchboards. The tabulator could then be made to tabulate different kinds of data merely by rearranging the plugs.[23]

In this sense, introducing a plugboard into the Hollerith tabulator systematized the process of redesigning the tabulator itself. Even though the plugboard did not *eliminate* the need to modify the physical structure of the tabulator to enable it to perform a new function, once you knew what kind of data you wanted to tabulate, rearranging the plugboard to enable the tabulator to crunch that data was straightforward. Introducing the plugboard transformed or *reduced* the complex problem of modifying the guts of the tabulator to the simpler problem of systematically rearranging cables on the plugboard. Furthermore, because every operation performed by the tabulator was then dictated by the particular configuration of cables on the plugboard, a particular programmed plugboard was a kind of software representing a program for tabulating that particular kind of data.

Electrifying Software

Although a punched card may be software, it is a rather unwieldy and limited kind of software. Because its holes need to be relatively big, it can hold only a small amount of information before the card itself becomes impractically large. Punching holes in it requires applying a large amount of force to it with a sharp pin. Although a machine can read information from the holes more quickly than a human can read the equivalent information from paper, the speed at which pins can be pushed through the holes imposes a limit on the speed of any card-reading machine. Finally, and perhaps worst of all, once you have punched holes in the card you can't erase or change them. Therefore, making even a minor modification to punchcard software requires punching a whole new set of cards.

It was with these limitations in mind that engineers in the 1940s developed magnetic and electronic, rather than mechanical, media for storing data and programs. For example, the U.S. census in 1950 used iron-plated metal tape instead of punched cards to store census data magnetically—and much more compactly—than was possible with paper, leading private companies to adopt such magnetic storage systems rapidly for their own purposes.[24]

Today we're all familiar with electrical and magnetic storage media in the form of hard disk drives, CDs, flash memories, and the random access memory (RAM) inside our computers. Although the technologies underlying these forms of storage vary widely, all of them store software in the form of electromagnetic signals rather than rope, punched holes, or plugs in a board. As a result, they can store more software in a smaller space, read and write such software more quickly, and both modify and erase existing software even while it is running.

None of these features of electromagnetic storage media, however, undermines the conclusion that even software stored as invisible electromagnetic signals in a modern computer's memory *physically* causes the computer to perform certain actions in just as real a way as rope wound in one direction around the Ropebot's axle causes it to roll backward. The physically "softest" software is still "hard" in this critical sense.

Software Hardens

Although the general historical trend *so far* has been for software to take increasingly "soft" physical forms over time, even today's software is often embodied in matter. For example, a CD stores software in patterns of physical depressions called "pits." (The parts that remain raised are called "lands.") Although a CD drive *reads* software from a CD using light, the software itself is stored in patterns of matter.

Furthermore, software already is showing signs of rehardening in other ways. For example, in the nascent field of DNA computing, human-created software is stored in strands of DNA, which then executes the software using biological processes. DNA's ability to store and execute software should not be surprising given the fact that DNA, the "code of life," is nature's way of physically storing instructions for creating biological organisms.[25] Although DNA computing and other matter-based technologies, such as molecular computing,[26] are in their infancy, researchers are pursuing them actively because of their potential to perform computations more quickly and in less space than is possible using electronic technology.[27]

Reframing the Hardware-Software Divide

Software can be embodied in matter, so the hardware-software distinction must rest on something other than the distinction between matter and energy.[28] Let me propose an alternative distinction: hardware is hard in the sense that it is the *fixed* part of a computer, whereas software is soft in the sense that it is the *variable* part of a computer. The hardware of the Ropebot consists of its axle, wheels, pins, and other components that remain unchanged *regard-*

less of how the Ropebot is programmed. The Ropebot's software is the rope wound round its axle in a particular configuration. Although at any point in time the rope is made of matter and has a particular fixed shape, what makes the rope software is that you can *change* the rope windings to change the path taken by the Ropebot.

What makes software soft, therefore, is not that the software *itself* is malleable, but rather that the actions performed by the *machine* the software controls can be changed merely by furnishing the machine with new software. This distinction holds whether a particular machine requires its software to take the form of matter—rope, punched cards, molecules, DNA—or energy, such as stored electrical or magnetic signals.

Programmable machines do nothing useful without software. In this sense, hardware is to software as a drill is to a drill bit. Turn on an electric drill without any bits attached, and the drill head will spin furiously but do nothing useful. The same is true of a computer without any software installed on it. This may run counter to your experience unless you are a computer geek, because most computers today are sold with software preinstalled on them, much like a drill sold with a bit already mounted to it. But the bare hardware of a computer is a clean slate, containing no software. Turn it on and it will do nothing but run up your electricity bill and warm your feet.

Attach a flat-head screwdriver bit to the head of a power drill and turn on the power, however, and you can drive screws with ease. Replace the screwdriver bit with a sanding attachment, and you can smooth surfaces to your heart's content. Each bit enables the drill to perform a function dictated by the physical structure of the bit.

The same is true of a computer. Start with a computer fresh out of the factory and install calculator software on it, and the computer becomes capable of performing arithmetic. Install word processing software, accounting software, or videogame software on the computer, and in each case you enable the computer to perform yet another task, each dictated by the software's physical interaction with the computer's hardware. In this sense, computer software is to computer hardware as a drill bit is to a drill.

No analogy is applicable in all circumstances. For example, it isn't particularly fruitful to think about digital music as a drill bit, even though technically such music is software in the sense that it instructs the computer's hardware to produce a particular sequence of sounds. A better analogy for digital music is a player piano roll or phonograph record. My purpose, however, is not to propose a perfect analogy—itself an oxymoron—but rather a framework that is more useful for thinking about precisely the class of software for which the drill analogy is most apt: software that performs functions similar to that of traditional machines, such as software for controlling the brakes of your car or for guiding a missile. Such software, if it is new and performs a practical function, has all the makings of an invention. If the process of creating such software has been automated, then the process is one of *automated inventing*. Adopting more common analogies, such as "software is to hardware as a recipe is to a chef," forecloses even the possibility that software could be an invention; no chocolate cake recipe, no matter how inspired, can ever qualify as an invention.

Although early programmable machines such as the Ropebot and Jacquard Loom could not be programmed to create new *inventions*, they nonetheless were nascent forms of artificial genies in the sense that their operation could be controlled using *instructions*. The essence of an artificial invention genie is that you can give it a wish describing the problem you want to solve, in response to which the genie automatically produces a design for a machine—or a physical machine itself—that solves the problem. Such invention-generating artificial genies would have to wait until programmable machines acquired a new kind of flexibility and an ability to both model and interact with the outside world.

From Twine to Telepathy

The Ropebot could not be programmed by typing a program into it using a keyboard and then typing RUN, as I was able to do with my TRS-80 BASIC program two millennia later. Even if Hero, the Ropebot's inventor, had planned out a course for the Ropebot by first writing down a sequence of instructions, to store those instructions in the Ropebot he had to physically wind rope in the right pattern around the Ropebot's axle.

Because performing such physical translation of human-readable instructions into patterns of rope is so unenviable a job, the modern history of programmable machines is replete with examples of efforts to automate such translation. Modern input devices, such as the seemingly humble keyboard, have played a key role in such automation. Just consider that to store a program on a Hollerith punchcard you had to insert a blank card into a hole-punching machine and move a mechanical pointer to positions corresponding to the holes to be punched.[29] In such early systems, there was little physical or conceptual distance separating the programmer's physical actions from the physical form in which the resulting software was stored.[30]

In contrast, consider even a simple one-line computer program written for a modern computer, such as "Add 2 + 4."[31] When a programmer uses a keyboard to type the text of such a program, the computer responds by automatically storing the instruction Add 2 + 4 in its memory as software consisting of a long sequence of bits, say 01101010000000001000000100.[32] The computer stores this bit sequence in memory by flipping a sequence of switches in an on-off pattern corresponding to the ones and zeros in the bit sequence.

In the programming example as I've described it, the computer *creates* software *within* the computer itself. The same basic process, however, can be used to create software and devices *external* to the computer. Burning software onto a CD is a simple example, but any software can also be burned onto devices such as read only memory (ROM), which is often used itself as hardware in a computer. More impressively, engineers who design today's microprocessors do so not by designing each of the processor's millions of transistors by hand but rather by writing *instructions* in a hardware description language (HDL). A software program called a compiler or synthesizer then automatically transforms those instructions into a schematic for the processor itself. This schematic can even be sent to a foundry to be manufactured automatically. Sound familiar? An abstract description of what the engineer wants the circuit to do is translated into a concrete description of components for doing it. The genie is at work again, as shown in Figure 3. The HDL description written by the engineer is like a wish, and the HDL synthesizer, which translates that wish into a design for a physical circuit, is like a genie.

The entire process is referred to as electronic design automation (EDA) and is now a mature field; its flagship event, the Design Automation Conference, dates back to 1964 and has an annual attendance of more than 10,000. Today's

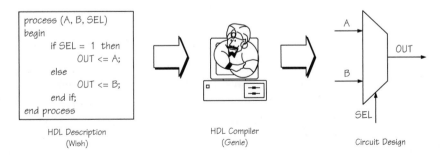

```
process (A, B, SEL)
begin
    if SEL = 1 then
        OUT <= A;
    else
        OUT <= B;
    end if;
end process
```

HDL Description   HDL Compiler
(Wish)   (Genie)   Circuit Design

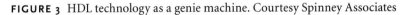

FIGURE 3 HDL technology as a genie machine. Courtesy Spinney Associates

processors, such as the latest Pentium or Athlon processor, are designed this way not simply out of convenience but by necessity. Back in 1971, Federico Faggin designed Intel's 4004 processor, with 2,300 transistors, manually—transistor by transistor.[33] As of this writing, Intel's latest processors have more than one billion transistors—far too many for any team of engineers to design using their unaided brains, no matter how much coffee they drink.

Keyboards are just one example of diverse physical input devices that have been used to ease the programmer's burden of funneling software into a computer's memory. Programmers, just like other computer users, use mice, trackballs, styluses, and microphones combined with speech recognition software to input their programs. Recent advances have even been made in controlling computers with thoughts—computer mind control. For example, a 14-year-old who suffers from epilepsy was able to play the classic videogame Space Invaders using only signals transmitted by a device attached directly to his brain.[34] Other brain-computer interfaces have been used to enable quadriplegics to perform point-and-click operations in a computer's graphical user interface using their brain alone.[35] If such technology ever advances to the point where humans can transmit instructions in a programming language directly to a computer, then programmers will be able to throw out their keyboards and program computers by telepathy.

What all of these mechanisms have in common is that they furnish humans with a more *human* vehicle for providing instructions to a computer. Over a half-century after the birth of the modern computer, programming a computer remains fundamentally an act of flipping the right switches in the computer's memory to make the computer do what you want it to do. What has changed is that computer scientists and engineers have developed input devices that increasingly enable programmers to use human-friendly physical actions to flip those switches. Such input devices are effective, however, only because modern computers are also equipped with mechanisms for *automatically transforming* the physical actions that programmers perform on those input devices—such as pressing a key or moving a mouse—into working software inside the computer. By greasing the wheels of translating human desires expressed in written language into physical structures for satisfying those desires, we have begun to imbue computers with some of the power of genies.

2 Enter the Universal Machine

A PROGRAMMABLE MACHINE, by its nature, can perform many processes, one for each possible program. A Ropebot can travel as many paths as there are ways to wind rope around its axle. As a result, a programmable machine has a greater degree of flexibility than many other kinds of machines. A toaster toasts and a hammer hammers; just try toasting bread with a hammer or driving nails with a toaster if you think otherwise. For this reason, toasters and hammers are both *special-purpose machines.*

Some devices, such as a combination TV-DVD-VCR or a Swiss Army Knife, can perform multiple functions. But no matter how many ways you can slice and dice with a Swiss Army Knife, you can't change its operation by programming it. Therefore, a Swiss Army Knife is a *multifunction machine,* but not a programmable machine.

Not all programmable machines, however, are artificial invention genies. A Jacquard Loom will never produce an *invention* no matter how cleverly you program it; it will only produce a cloth with a new pattern, which we would not consider to be an invention no matter how new it might be.[1]

Computers Go Universal

A modern computer, unlike a Jacquard Loom, can be programmed to perform a nearly infinite number of functions. In fact, you could program a single computer to act as the guts of a Jacquard Loom *and* a Hollerith tabulator simultaneously. You are familiar with the extreme flexibility of modern computers just from using them. The computer I am using to write this book has software installed on it for writing documents, browsing the World Wide Web, compressing and decompressing files, playing music and video, and protecting itself against viruses, just to name a few. Furthermore, I can install new software on the computer to enable it to perform new functions whenever I please, even if the designers of the computer itself never dreamed of those functions. I can even run multiple pieces of software simultaneously on the same computer, as you know if you've ever checked your email in one window while browsing the Web in another. Try doing all of that with a Hollerith tabulator.

Modern computers are often called *general-purpose machines* or *universal machines* as a result of this shape-shifting ability.[2] Recall the analogy to drill and drill bit, but now take it one step further. Attaching a sanding bit to a drill doesn't just enable the drill to sand surfaces; it effectively *transforms* the drill into a power sander. The same is true of a computer. A computer with no software installed on it is a universal machine, a blank slate pregnant with possibilities, none of which has yet been unlocked. Install word processing software on the computer, however, and you effectively transform the general-purpose computer into a special-purpose machine for writing documents.

The term *universal machine* derives historically from Alan Turing's use of it in a seminal 1936 paper[3] in which he demonstrated that a longstanding problem in abstract mathematics posed by mathematician David Hilbert at the turn of the 20th century could not be solved.[4] Incredibly, in retrospect, Turing described in principle how a universal machine could work merely for the purpose of proving his primary mathematical conclusion. Turing, in other words, laid the theoretical foundation for all modern computers as part of an intellectual detour.

Although Turing's original concept of the universal machine was of an *abstract* machine, not a physical one, it wasn't long before he and others recognized that a real, physical, universal machine could be built. Although such pioneers built a variety of machines with increasing degrees of programmability in the years after Turing's original paper was published, it was not until 1945, when John von Neumann formulated a feasible way to implement a "stored program computer" in a real machine,[5] that the dream of building working universal machines became a reality.[6]

A computer designed according to von Neumann's stored program architecture consists of a processor and a main memory connected by a "bus" (one or more wires for shuttling information between the processor and the memory), as shown in Figure 4. The primary novelty of von Neumann's design lay in the fact that you could store an entire computer program and the data on which it operates—what we call software today—in a single uniform memory. By comparison, recall the Hollerith tabulator with a plugboard, in which the *instructions* were embodied in the wired plugboard, while the *data* to be processed by the machine were embodied in a distinct set of punched cards.[7] Now consider a program with just a single line of code ("Add 2 + 4") in which Add is the *instruction* and the numbers 2 and 4 are the *data*. In a von Neumann computer, such a program could be represented in a single sequence of bits in memory, such as 01101010000000001000000100.

One significant benefit of this design was that the program instructions and data could be modified *while the program was running*, thereby opening the door to much more flexible and powerful programs than were possible with previous generations of computers.[8] Furthermore, making it possible to program a computer merely by storing software inside the computer's memory eliminated the need to engage in the tedious rewiring that was necessary to program early computers.[9] The von Neumann architecture has remained the basis of most modern computers ever since; every computer you have ever used most likely has been a von Neumann computer.

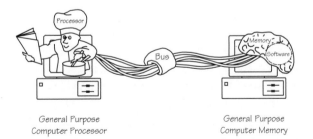

General Purpose
Computer Processor

General Purpose
Computer Memory

FIGURE 4 Programming a von Neumann computer.
Courtesy Spinney Associates

The Universal Machine as Artificial Invention Genie

The universal machine, as pioneers such as Turing and von Neumann origi-nally envisioned it, was capable of performing any *computation*. It probably isn't immediately obvious, however, how you could ever use such a souped-up calculator to produce an *invention* no matter how cleverly you might program it. You can perform computations using a handheld calculator until you're blue in the face and it will never morph into a new automobile engine or cancer-curing drug.

To understand how programming a universal machine can produce an invention, consider a thought experiment. Imagine that I show you two auto-mobiles and tell you that each has been equipped with a new and improved antilock braking system that has been proven to reduce skidding more effec-tively than any previous system. You test-drive the first car by driving it at 100 miles per hour and then slamming on the brakes. As advertised, the car slows down rapidly but smoothly to a stop without skidding or injuring you. Then you test-drive the second car in exactly the same way, and experience exactly the same results. You continue to test both cars under various conditions—turning the wheel while braking, braking while on ice, and so on—and in every case both cars operate indistinguishably from each other.

Now I open the hood of both cars and point to a black box inside each one. The black boxes contain the cars' respective antilock braking systems. You can see cables coming from the brake pedal of each car to the black box, and lead-ing from the black box to the brakes. When you press the brake pedal, a signal is sent from the brake pedal to the black box. Something inside the black box determines how much force to apply to the brakes, and how frequently to ap-ply that force, on the basis of the amount of force you apply to the brake pedal and other factors, such as the current speed and direction of the car. The an-tilock braking system in the black box, in other words, *controls* the brakes.

If you take my word for it that the antilock braking systems in both cars are in some way both *different* and *better* than any previous antilock braking system, would you agree that whatever is inside each black box is an inven-tion? At the very least, would you agree that either *both* black boxes are in-ventions or *neither* of them is, and that it cannot be true that one of the black boxes is an invention while the other is not?

You cannot stand the suspense any longer, so you open both black boxes and peer inside. In the first box you find a complex jumble of circuitry, which was custom-designed by an electrical engineer. In the other box you find a small laptop computer running antilock braking software written by a computer programmer. The circuitry in the first box and the software in the second box perform exactly the same sequence of steps to control their respective brakes. They just perform those steps using different physical mechanisms.

One motivation for this hypothetical scenario is to reinforce the conclusion that software is not "intangible" or a mere abstraction in any sense that is relevant to whether it is an invention. The software in the second antilock braking system takes a physical form, physically controlling the second car's brakes just as much as the first car's circuitry-based antilock braking system controls its brakes. If you deny this, then you shouldn't mind if I erase the antilock braking software from the second car and have you drive in it toward a cliff at high speed with only the antilock braking system to avert your fall. Although it may be possible to *describe* the software-based system using an abstract sequence of steps represented by a computer program, the same description describes the circuitry-based system just as accurately. The fact that you can describe a physical object in abstract terms does not make the object *itself* abstract. A comfy physical chair is still a physical chair even if I can describe it abstractly as "a device for stimulating thought" (or sleep).[10]

Another motivation for this thought experiment is to demonstrate that the act of programming a computer can produce a programmed computer that qualifies as an invention according to the same criteria as more traditional inventions, namely that it is new and useful. If the circuitry-based antilock braking system is an invention because of its improved antiskidding ability, then the software-based system is an invention for the same reason. At the very least, if the software-based antilock braking system is *not* an invention, then this must be so for some reason other than that the software is not new or useful.[11]

Although you can *think* of what the antilock braking software does as merely performing a sequence of computations, what allows the software to bridge the gap from mere computation to invention is that the computer on which the software executes is *connected to and physically controls* an external device to perform a useful function, which, when performed by a non-software device, qualifies that device as an invention.[12] Many inventions are useful precisely because of their ability to control other machines, or other components of the same machine; thermostats, automobile engines, and gears come to mind.[13] We don't rule out new and useful instances of such devices as inventions simply because their purpose is to act as an intermediary for controlling another device, rather than to perform a function that is directly useful to a human. Manufacturers of everything from radar systems to microwave ovens to wristwatches have been replacing mechanical and electrical components with programmed microprocessors for decades.[14] It seems odd at best to conclude that such manufacturers stopped inventing on the day they switched from electromechanical components to software-based components for performing the same functions.

Now put yourself in the shoes of the programmer who wrote the antilock braking software in the second car. You wrote a program describing a new antilock braking method and loaded that program into a computer, which created antilock braking software in its memory. By programming the computer, you transformed it into a different machine, or at least into a modified and improved version of its former self. If the transformed computer is new and useful, then it is an invention. Therefore, you produced an invention— the modified computer—by programming the computer. I call this general process "inventing by programming," which is the first historical example of "inventing by wishing."

You wrote the antilock braking program in a human-readable language to express the *actions* you wanted the computer to perform. For example, your program may have included instructions such as: "As the driver further depresses the brake pedal, increase the force applied to the brakes," and "Depress and release the brakes every half-second" (expressed much more precisely in a computer programming language, of course).

Such instructions are notable for what they do *not* specify: namely, which *physical mechanisms*, such as wires or pistons, to use to apply force to the brakes or to receive information about how much pressure the driver is applying to the brake pedal. Compare this to the kind of circuit schematic that would have been drawn by an electrical engineer who designed the antilock braking system in the other car. The schematic diagram would depict the physical circuit components to include in the antilock braking system to receive signals from the brake pedal and to transmit signals to the brakes themselves. Your code says nothing about these physical details. Instead, your program describes *what* you want the software to do, rather than *how* to do it physically. Programs, in other words, typically describe logical *functions*, not physical *structures*.[15] It is the computer that automatically created the necessary structure— in the form of physical software—according to the functional description you gave it. The computer, in other words, *automatically* transformed your written description of what you wanted the software to *do* into the physical structure necessary to do it. Because the resulting software is an invention, the process of creating the physical structure of that software automatically is a kind of *automated inventing*. Computers in this sense fit the definition of an artificial invention genie, which is why I call the traditional process of computer programming an instance of Genies v1.0.

Genies in Hiding

Computers are the ultimate artificial chameleons. We use them every day to mimic everything from old-fashioned typewriters to wristwatches to stereo systems. Although we now take the ubiquity and flexibility of computers for granted, most early computer engineers did not envision that computers would ever be so broadly applicable. Even into the 1960s, many thought that computers were useful solely for computing ballistics tables and performing other "data processing" tasks for large organizations and would never find any use outside this narrow range of functions.[16] This impoverished vision of computers as merely high-powered calculators resulted in part from focusing excessively on the *internal* flexibility of computers to perform any computation, while ignoring the potential of computers to interact *externally* with other machines and even directly with humans. Even Howard Aiken, one of the founders of modern computing, expressed his skepticism in 1956 that computers would ever be able to mimic a wide variety of machines when he said, "If it should turn out that the basic logics of a machine designed for the numerical solution of differential equations coincide with the logics of a machine intended to make bills for a department store, I would regard this as the most amazing coincidence that I have ever encountered."[17]

Such myopia has been reinforced, unfortunately, by the very name *computer*, which implies that such a machine is limited to performing calculations and other operations that are somehow *equivalent* to performing calculations. Some computing pioneers, however, recognized early on that computers had the potential to be much more than high-speed calculators. The team that designed ENIAC, one of the first computers, in the early 1940s "realized that what they were constructing would not only become the ultimate mathematical calculator, but the first, necessarily imperfect prototype of a whole new category of machine."[18]

The ENIAC team, however, was scooped more than a century earlier by Ada Byron, Countess of Lovelace, daughter of Lord Byron.[19] Lady Lovelace,[20] a skilled mathematician, was impressed with Charles Babbage's 1837 design for a programmable machine that he called the Analytical Engine. Babbage realized, decades before anyone else, that a multitude of machines for performing diverse calculations could be replaced with a single machine to perform any of those calculations.[21] His design for the Analytical Engine achieved this result by incorporating mechanisms for changing the order in which calculations were performed, in response to instructions and numbers provided to the machine on modified versions of Jacquard's punched cards.[22] As Ada put it, "the Analytical Engine *weaves algebraical patterns* just as the Jacquard-loom weaves flowers and leaves."[23]

Ada Byron recognized that the Analytical Engine was not a mere calculating machine but rather a fundamentally new kind of machine whose programs could be represented by written instructions. In fact, it has been said that Ada "fully understood the principles of a programmed computer a century before its time."[24] The Ada programming language was named after her in recognition of her contributions to the foundations of programming.

The longstanding dominance of the narrower view of computers as high-speed programmable calculators helps to explain in part why the role of computers as invention-automating machines has been overlooked for so long. If computers are merely calculators, then no program—and no computer that has been programmed with that program—would ever seem to qualify as an invention merely by virtue of the contents of the program itself. If a new computer were to be an invention, it would have to be due to some unique feature of its hardware rather than the software stored in it.

Turing and others who recognized that computers could do much more than compute did so not primarily as the result of creating more flexible computer hardware but by expanding how they *thought* about the nature of computers. Once they realized that the internal operations performed by a computer could be conceptualized not just as calculating numerical quantities but also as manipulating images, sounds, and other representations of the world *outside* the computer (such as the force applied to a car's brakes), they opened their minds to the nearly infinite real-world applications of computers that we enjoy today, including the ability to use programmed computers as *substitutes* for other kinds of machines in increasingly varied and powerful ways.[25]

3   The Rise of Wishes

SO FAR I'VE REFERRED TO PROGRAMMING as the *physical act* of creating software. This is primarily what you mean when you say that you have just programmed your DVR to record an upcoming television show, even though your spouse told you which show to record. You did not decide which shows to schedule or exercise any other discretion; you simply carried out the physical actions necessary to configure the DVR to record the shows that someone else selected.

We turn now to another meaning of programming: the act of figuring out which instructions to include in the program in the first place. You can engage in this kind of programming by writing instructions on paper, or even forming them in your mind. Real programmers often do this by sketching out a program in pseudocode, a mixture of a real programming language and English that is easier to write quickly, before writing the real program to be stored in a computer.

Building Towers of Abstraction

The processor in most computers is hardwired to understand a relatively small number of primitive instructions, such as "Add," "Subtract," "Multiply," "Divide," and instructions for making decisions based on the outcome of previous instructions. The set of instructions that a particular processor understands, combined with the permissible rules for combining those instructions together into programs, is referred to as the processor's "machine language."

Most programmers today, however, do not write software in the native tongue of a computer processor. Instead, they use "high-level" programming languages that include much more abstract and powerful instructions. A single instruction in a high-level programming language may, for example, be capable of sending an email message over the Internet or even displaying an animated three-dimensional character in a videogame—a far cry from "Add" and "Subtract."

An instruction such as "send the following email message to abc@xyz.com" does not derive its power from more advanced hardware. Instead, when a programmer writes such a "send email" instruction, the computer translates that *source code* instruction into *object code* containing many—possibly thousands—of low-level machine-language instructions that, when performed together, send an email message. Typically, when you purchase email software such as Qualcomm Eudora or Microsoft Outlook, it contains the object code, ready to run, rather than the high-level source code that the programmer originally wrote.

This only begs the question: How does the computer know which machine-language instructions to substitute for the single high-level "send email" instruction? The short answer is that the entire history of computer science can be seen, in one sense, as an effort to enable computers to translate increasingly abstract and powerful instructions written in increasingly high-level programming languages into the necessary low-level machine-language instructions.

Before we examine this history, it is useful to clarify the meaning of the words *abstract* and *abstraction* as computer scientists use them.[1] Although sometimes people use the former term in common parlance to refer to concepts that are vague or subject to interpretation, nothing could be further from the meaning of abstract in computer science. To a computer scientist, an abstraction is a single concept that encompasses many specific instances. The concept "dog" is an abstraction of the three specific dogs on my street; "mammal" is an abstraction of the categories of dog, cat, and human; and "animal" is an abstraction of the categories of mammal, reptile, and amphibian. Abstractions often form hierarchical structures, as in the case of the biological taxonomy of life: the "root" is all of life, which branches out into domains, kingdoms, phyla, classes, and so on. One concept is more abstract than another concept in the same branch of the hierarchy if the first concept encompasses the second: the class mammal is more abstract than the order carnivore or the species *Canis lupus* within it.

Finally, computer scientists use the term *abstract* as a verb to refer to the process of *creating* abstractions. The process of recognizing that dogs, cats, and humans have something in common, and attaching the label "mammal" to that commonality, is an act of *abstracting* from three species to the more abstract class mammal.

There is nothing necessarily fuzzy about such abstractions or the process of creating them. As computer scientist Edsger W. Dijkstra said, "the purpose of abstracting is *not* to be vague, but to create a new semantic level in which one can be absolutely precise."[2] A mathematical equation, such as $y = mx + b$, is highly abstract in that it encompasses a general relation among an infinite set of numbers, but it is also "absolutely precise" in Dijkstra's terms.

The process of creating abstractions in computer science is similar to the process of abstracting from specific dogs to the abstract concept of dog, but with a twist. Nouns such as dog primarily describe *objects*. Since computer programs typically consist of instructions to a computer to perform particular *actions*, abstractions in programs may be analogized more fruitfully to abstractions in other written works that convey instructions, such as a cookbook containing recipes. Imagine a cookbook that assumes you, the reader, know absolutely nothing about how to cook—not even how to boil water. Then assume that the cookbook author wants you to be able to read any individual recipe in the cookbook and follow it straight through without having to read any other part of the book. The recipes in such a cookbook will be quite long. Every recipe that calls for boiling water will spell out the entire procedure ("get a pot, fill it with water, place it on the stovetop, turn on the burner," etc.) in great detail. Recipes that required boiling water multiple times will spell out the complete procedure each time.

A more rational approach is to define certain common cooking procedures up front: boiling water, toasting bread, cracking an egg, chopping onions. Then a particular recipe that calls for one of these procedures (a subroutine) will simply instruct you to "boil water as described on page 4." Such a statement concisely and abstractly captures, in a single sentence, the entire procedure for boiling water that has been defined previously.

The same process of encapsulating multiple instructions within predefined subroutines can be repeated. Once the cookbook provides a recipe for making scrambled eggs with onions (by reference to the egg-cracking and onion-chopping procedures), the cookbook can offer a concise "breakfast" recipe by instructing you to toast bread using the toasting procedure and make eggs using the scrambled egg recipe. The breakfast recipe is very abstract and brief—just two sentences—but requires you to perform many specific actions to carry it out. Furthermore, such a recipe represents an abstraction that forms a hierarchy, with the egg-cracking and onion-chopping procedures one level below it, and the specific actions for carrying out those procedures one level lower still.

Computer scientists build increasingly abstract and powerful programs and programming languages analogously. If you have only a bare-bones computer at your disposal, then you must write programs for it in the computer's native machine language. Machine-language instructions, however, are typically expressed as decimal or binary numbers, which are terribly difficult for human programmers to read and write. An "Add" instruction, for example, might be represented in machine language by decimal 132, which is 10000100 in binary. A complete instruction for adding 2 + 4 might therefore be represented in binary as 100001001000001010000100. Just try writing more than a few such instructions in a row and your mind will quickly grow numb. Therefore, programmers typically write such programs in a slightly more understandable language called an "assembly language," in which a human-readable mnemonic takes the place of each machine-language instruction. For example, the mnemonic "Add," also called an "op code," might stand in for 10000100.

If you want to add 2 + 4, you can do so by writing a single instruction: "Add 2 + 4." If, however, you want to perform a calculation, such as deriving a square root, that cannot be performed by any single instruction the processor is designed to understand, then you need to write a sequence of instructions that, in combination, perform the calculation you desire. Take my word for it that it is possible to write a sequence of primitive instructions consisting mostly of simple arithmetic operations to calculate the square root of any given number.

Now assume that you've written such a program, built out of 50 primitive instructions, to calculate a square root.

Just as with the recipe for boiling water, you do not need to rewrite the square root program (called a subroutine) over again each time you want to calculate a square root. Instead, you can just write the square root subroutine *once*, and then refer back to it by name whenever you want to calculate a square root. It is as if you have added a new, more abstract, SquareRoot instruction to the programming language you are using. For example, you could write a program that calculates the square root of every number from one through five as follows:

```
SquareRoot  1
SquareRoot  2
SquareRoot  3
SquareRoot  4
SquareRoot  5
```

Each line of this program is like the instruction in the cookbook recipe that says "Boil water using the procedure described on page 4." There is a key difference, however, between the cookbook example and programming: *a human still needs to carry out each step in the water boiling procedure manually to follow the recipe.* Writing "boil water" as step 2 in a recipe, rather than writing out the steps involved in boiling water, may save you ink, but it does not save whoever follows the recipe any time or effort. In contrast, the author of a *computer program* who writes a square root subroutine and then merely references it by name later *does* reduce the total amount of human effort that must be expended to execute the program because it is the *computer* that follows the recipe, automatically and obediently carrying out every step in the SquareRoot subroutine each time it encounters a SquareRoot instruction in the program. In other words, although we often define abstractions in cookbooks to save space, programmers define abstractions in programs both to save space *and* to enable computers to transform those abstractions into actions automatically.

Just as you can build procedures on top of procedures in a recipe book, you can build abstractions on top of abstractions in a computer program.[3] If you've just written a subroutine called "Line" for drawing a straight line on the screen, you can write a higher-level subroutine called "Square" for drawing a square by combining together four Line instructions. There is no limit to the number of *levels of abstraction* you can build in this way. If you could see the "source code" of just about any software on your computer today, you would easily find instructions that are built on ten or more nested levels of abstraction.[4]

Automating Translation

Programmers wrote programs in assembly language even before computers were capable of translating those instructions into machine language automatically, simply because it was easier to write programs in assembly language and then to translate those instructions manually into machine language than it was to write programs directly in machine language. This translation process is known as *assembly*, and the people who performed such mind-numbingly tedious translation were known as *assemblers*. Eventually people had the bright idea of *automating* the assembly process, using software that is called an "assembler" in homage to the human assemblers it replaced (none of whom uttered so much as a peep in protest). Similarly, the term *computer* itself originally referred to a *person* who carried out the instructions in a program.[5] Machine computers were referred to as automatic computers until the pervasiveness of the machine variety caused the qualifier "automatic" to be dropped from the name.[6]

As small as the gap may have been between writing assembly-language programs and physically flipping switches in the computer's memory, the introduction of automatic assemblers brought about the first true transition from programming by physical redesign to programming *solely* by writing instructions in a human-readable language, thereby bringing us one step closer to a true genie in a machine.

Stacking Automation on Automation

The problem with such early genies, however, was that they were not only relatively weak but also highly demanding; they could grant only wishes written in the very limited terms that could be expressed in assembly languages. Assembly-language programmers soon began to make the same complaints as the previous generation of programmers: writing programs in assembly language is tedious and time-consuming, and there is a limit to how complex we can make our programs using assembly language.[7] Writing assembly-language programs that could perform even relatively simple operations by today's standards, such as printing text on a printer, or storing data on a hard disk, required writing hundreds or even thousands of assembly language instructions.

We've already seen one way that programmers can tackle this problem: by writing subroutines (such as SquareRoot) to encapsulate commonly needed functions, and then using those subroutines as individual instructions themselves. Although this solution is an improvement over having to write every program from scratch in assembly language, it is as limiting as trying to write an article in the *New Yorker* by starting with a first-grade vocabulary and then expressing more abstract concepts only after you have first defined them in the article itself using the first-grade vocabulary. Writing for an adult audience is much easier if you can just use abstract terms right out of the gate.

Computer scientists developed early high-level programming languages, such as FORTRAN, COBOL, PASCAL, C, and BASIC, to address this limitation of assembly languages.[8] The "primitive" instructions in a high-level language are already more abstract and powerful than assembly-language instructions. For example, in a typical high-level language you can write an instruction to perform arithmetic in much the same way as you would write it on paper, such as $x = 3 + (2 \times 7) - (15/12)$, instead of writing out each individual subcalculation as a separate instruction. High-level languages include instructions for performing functions such as displaying text and graphics on a monitor, accessing information on a hard disk drive, and obtaining input from the user through a keyboard or mouse.

Furthermore, the grammar or *syntax* of an assembly language is limited essentially to verb-noun sentences; "Add 2 + 4" is a good example. High-level languages allow you to write sentences (statements) that resemble English more closely than assembly language statements, in forms such as "if A is true, then perform instruction B; otherwise, perform instruction C." These and other features of high-level languages facilitate writing powerful programs more quickly and easily than in assembly language by enabling programmers to focus on the high-level actions they want their programs to perform, rather than on the low-level details of how a computer processor is to perform those actions.

You can't just write a program in a high-level language on any computer and expect the computer to run the program; remember that a computer's processor understands only the machine language that has been hardwired into it. To make a computer run a program written in a high-level language, you need a program called a *compiler*, which, like an assembler, translates high-level programming-language instructions into machine-language instructions.[9]

Just as an individual programmer can build higher-and-higher-level instructions out of lower-level instructions, thereby producing increasingly abstract programs over time, so too have high-level programming languages evolved over time. The C programming language, once considered a high-level language, is now considered relatively low-level in comparison to more modern languages consisting of instructions even further removed from the direct physical operation of the computer processor.[10]

Even a high-level programming language may lack instructions for performing complex tasks, such as drawing three-dimensional graphics for use in a videogame. If you are designing such a game and want to avoid writing the graphics-drawing portion of the program from scratch yourself, chances are that you can buy or download a free *library* of software subroutines that someone else has already created for the purpose of easily incorporating those subroutines into larger programs.[11] This usage of the term *library* dates back at least to the Jacquard Loom, when weavers would store collections of cards for weaving commonly used patterns such as roses or borders in a physical library and retrieve those card collections for use as part of larger designs.[12]

This pattern of computer scientists and programmers using existing programming languages and automatic translation technology as a foundation for developing yet higher-level (more abstract) programming languages and translation technology has repeated many times throughout the history of computer science, leading to increasingly abstract programming languages and programs.[13] TenFold Corporation has taken this logical leapfrogging process to the next level by creating software that enables businesspeople with no training in computer programming to create business software applications for their companies without writing a line of code. TenFold gives its customers software modules that perform all of the functions a business software application might need, such as sending and receiving email, ensuring security, and sharing documents over a network—like Lego bricks that can be pieced together to form software applications. As a businessperson uses TenFold's software, he or she is led through a detailed series of questions about what the software must do. Do you need email? Password protection? Payroll processing? For each topic, the software drills down to low-level details such as the minimum number of characters required for a password and how many times users should be allowed to enter incorrect passwords before being locked out of the system. Then, from the user's answers, TenFold's software builds a *description* of the software that the businessperson wants to create and "renders" that description into a business application that performs all of the specified functions by configuring and stitching together the necessary precoded building blocks—all without requiring the businessperson to write a stitch of code, or even to know how to program. Sound familiar? The genie rears its head once again. According to TenFold President Jeff Walker, the company's patented[14] process can perform in a few months what previously often required two hundred programmers five years to accomplish.[15]

Similar growth in productivity resulting from increasing abstraction and automation has been observed since the inception of high-level programming languages. When John Backus proposed the first high-level programming language, FORTRAN, in 1954, he "promised a system that would generate [machine] code as good as the code that could be produced by a human [assembly language] programmer. Most programmers and their managers were highly skeptical" that a machine translation of high-level instructions into machine code would produce machine-language code that could compete directly with human assembly-language programmers. When Backus finally produced a working FORTRAN system in 1957,

> the responses from users [were] immediate and ecstatic. General Motors estimated that the productivity of programmers was increased by a factor of between 5 and 10, and that, even with the machine time consumed by the compiler taken into account, the overall cost of programming was reduced by a factor of 2.5.[16]

Similarly, the advent of hardware description languages for creating integrated circuits by writing descriptions of them immediately brought about a hundredfold increase in the complexity of chips that could be designed.[17]

The shift from old-fashioned inventing to inventing by programming in such a wide range of contexts should come as no surprise in light of these benefits. Although one reason for the shift is that microprocessors programmed with software often are less expensive to *manufacture* than custom-designed components that perform the same function, an equally important reason is that it is simpler and less expensive to *design* such software. Although engineers a hundred years ago knew how to manage the complexity of the design process by decomposing a complex problem into smaller, more easily soluble, problems, such decomposition still required the engineers to (1) design *physical components* that could solve each subproblem and (2) *physically construct* those components. Performing the first step is, in large part, what we traditionally have referred to as inventing. For example, in the time leading up to Samuel Morse's invention of the telegraph, he and other engineers of the day strongly suspected that a device could be built to harness electricity and transmit messages over essentially unlimited distances.[18] They even had a pretty good idea *in general* of which kinds of components would be needed to do the job. What they *lacked* was a concrete physical design for such a machine, and Morse was the first to crack the nut when he designed the physical structure of the telegraph. This is why we refer to him as the telegraph's "inventor."

Enabling an engineer to design and construct a working machine merely by writing a description of the functions it is to perform greatly simplifies the engineer's job by *eliminating* the need to take the extra, historically crucial, steps of designing the detailed physical structure of the machine and then building it. The lure and benefits of such automation are precisely what has caused so many individuals and companies, in fields ranging from radar systems to anti-lock brakes to microwave ovens, to switch over to the "inventing by programming" camp as technology continues to enable them to create wider and wider varieties of machines simply by writing abstract descriptions of them.

The Mentalist Fallacy

Programmers today typically spend their day writing abstract descriptions of what they want computers to do, and leaving it to computers to create software to do it. Such personal detachment from the nuts and bolts of computer technology on the part of the very people who create software carries with it a risk that I call the Mentalist Fallacy: the mistaken belief that because you can *create* software merely by writing abstract instructions, the resulting software *is* nothing more than abstract instructions, lacking any physical manifestation in the real world. As we will see in more detail when we review the history of software patents, the Mentalist Fallacy has caused unfortunate and unnecessary confusion about how to apply patent law to software.

If I am right that those who are most well-versed in programming are the most susceptible to the Mentalist Fallacy, then you would expect accomplished programmers and computer scientists to fall prey to a particularly strong form of it. Instances of exactly this are well documented. For example, Richard Stallman, creator of the legendary (among computer geeks) emacs text editing software and the progenitor of the concept of "free software," stated in testimony to the U.S. Patent Office that "software is like other fields of engineering in many ways. But there is a fundamental difference: computer programs are built out of ideal mathematical objects."[19] Jim Warren, founding editor of the computer software trade magazine *Dr. Dobb's Journal* and former board member of computer-aided design (CAD) company Autodesk, similarly testified before the U.S. Congress that "software is not a gadget. . . . Software is what occurs between stimulus and response, with no physical incarnation other than as representations of binary logic."[20] Even the legendary computer science professor Donald Knuth stated his belief, in a letter to the Patent Office, that "the Patent Office has fulfilled this mission [of serving society by formulating patent law] with respect to aspects of technology that involve concrete laws of physics rather than abstract laws of thought," such as software.[21]

A similarly flawed, and equally common, claim is that software merely carries out "mental" processes.[22] But it is a mistake to conclude that software *is* a mental process merely because it *mimics* or *replaces* mental processes, for the same reason it would be a mistake to conclude that a pocket calculator *is* a mental process merely because it mimics or replaces the process of performing arithmetic mentally. Adding 2 + 2 is a "mental" operation only when, by definition, you perform that operation using your mind. When an electronic calculator adds the same two numbers, it is by definition performing an electronic operation, not a mental one, since the calculator does not have a mind. Calculation, in other words, is a mental process when it is carried out by a mind but not when it is carried out by a device lacking a mind. It seems that we still have difficulty accepting that our mental activities can be mimicked by machines—almost four centuries after Blaise Pascal invented a mechanical calculating device that should have "mark[e]d the final break with the long age of ignorance, superstition and mysticism which above all had stopped the human race from contemplating that certain mental operations could be consigned to material structures made up of mechanical elements, designed to obtain the same results."[23]

Another lure of the Mentalist Fallacy is that one of the most common metaphors for the operation of a computer—a metaphor that I admittedly use throughout this book—is that the computer "reads" and "understands" "instructions"—all analogies to the operation of the human mind. Although this metaphor is useful in many circumstances, overreliance on it to the exclusion of other metaphors (such as drill and drill bit) can reinforce the conclusion that nothing physical occurs when a computer executes software, particularly if you are prone to accepting the Cartesian mind-body dichotomy. Although it may be useful to think of and talk about a computer as if it understands instructions in the same way as a human mind, in fact the computer does not understand anything. The computer's processor simply reacts physically to the physical software stored in its memory, just as a phonograph reacts physically to the physical grooves in a phonograph record.[24]

No matter how much we automate the mechanisms for *transforming* human-written instructions into working software, and no matter how much we hide the inner workings of that transformation process from the view of computer programmers and computer users, the end result is the same, whether we're talking about the Ropebot, the Jacquard Loom, or a cell phone with Web browsing software installed on it—a machine whose physical structure has been modified to make it capable of doing what a human programmer wanted it to do. R. W. Hamming, a recipient of the Association for Computing Machinery's (ACM) prestigious Turing Award, recognized the temptation of the Mentalist Fallacy as long ago as 1969, and in response to that temptation said:

> At the heart of computer science lies a technological device, the computing machine. Without the machine almost all of what we do would become idle speculation, hardly different from that of the notorious Scholastics of the Middle Ages. The founders of the ACM clearly recognized that most of what we did, or were going to do, rested on this technological device, and they deliberately included the word "machinery" in the title [of the ACM]. There are those who would like to eliminate the word, in a sense to symbolically free the field from reality, but so far these efforts have failed. I do not regret the initial choice. I still believe that it is important for us to recognize that the computer, the information processing machine, is the foundation of our field.[25]

Although we may be able to make our wishes come true in the Artificial Invention Age merely by writing those wishes, we must never forget that wishes are powerless without genies to grant them, and that once granted the wish-come-true is just as real and tangible as if we had built it by hand.

Relieving Invention Fatigue

I don't think necessity is the mother of invention—invention, in my opinion, arises directly from idleness, possibly also from laziness. To save oneself trouble.

—*Agatha Christie*[26]

Programmers were by no means the first class of workers to desire machines that could do their dirty work for them, thereby freeing them to engage in loftier pursuits. Mathematicians and their professional counterparts—accountants, bookkeepers, and actuaries—have long been plagued by the grueling tedium of days filled with unending manual calculation.[27] William Seward Burroughs felt compelled to leave his career as a bank clerk to escape the endless streams of numbers that filled not only his waking thoughts but his nightly dreams as well. As a result, he invented a mechanical adding machine that was the basis of the Burroughs Adding Machine Company in 1885.[28]

Astronomers are another bunch whose work has involved mind-numbing calculations for centuries. Predicting the paths of the stars requires computing logarithms, which are notoriously difficult to calculate by hand. As an example, the logarithm of 100 with a "base" of 10 is 2, because 10 to the power of 2 is equal to 100. Although the answer in this example is relatively easy to find because we already know that $10^2 = 100$, just try finding the logarithm of 3,428 without a calculator. (The answer: approximately 3.535.) As a result, astronomers throughout the ages have generated books filled with tables listing the logarithms of hundreds or thousands of numbers. Then, when they needed to identify the logarithm of a particular number, they would look up the answer in a preprinted logarithm table. Although this was time-consuming, it was not nearly as grueling as calculating the result from scratch every time.

Charles Babbage, who we met earlier in conjunction with his 19th century Analytical Engine, was both a mathematician and an astronomer. Before he designed the Engine, the tedium of calculating and reading the notoriously unreliable logarithmic tables that were necessary to perform the astronomy of the day prompted him to wonder whether the task of generating such tables could be automated by machine.[29] Babbage subsequently designed and constructed such a machine using cogs, wheels, and shafts.[30]

Mission: Eliminate Drudgery

What quantities of precious observations have been of no use to the advancement of science and technology, for the reason that there were not the resources needed to calculate results from them! How the prospects of long and arid calculation have demoralised great thinkers, who seek only time to meditate but instead see themselves swamped by the sheer mass of arithmetic to be done by an inadequate system!

—L. F. Menabrea[31]

The most dedicated and successful proponent of the use of computers to relieve drudgery was J.C.R. Licklider, who began his career as a psychoacoustics researcher with no formal training in computer science or engineering. Licklider, frustrated at how much time he seemed to be spending on routine administrative tasks rather than grappling with scientific problems, decided to apply his scientific training to himself by tracking how he spent every minute of his work day. He was surprised to find that he spent about 85 percent of his time performing clerical tasks and very little time engaged in the core intellectual work of "real" science. He concluded that most of the work he, and other scientists, actually performed would be done better by machines acting as "electronic file clerks," performing routine clerical tasks automatically, thereby freeing scientists to spend a greater percentage of their time engaged in higher-level tasks.[32]

Although Licklider initially focused on the liberating impact of such automation on scientists, the same principle applies to engineers and other inventors in their daily work designing new machines. Inventors have been dragged down by the tedium of their work just as much as mathematicians, astronomers, and other scientists. Designing a resilient bridge requires performing calculations based on the laws of physics. Supplying a structural engineer with a machine for automatically performing such calculations frees the engineer to focus on the overall shape, size, appearance, and usefulness of the bridge and how it can best serve human needs.

As we already have seen, computer scientists have become expert at automating the transformation of low-level abstractions into working software and then building on the resulting software to enable the automatic transformation of even higher-level abstractions into working software—with no end to such leapfrogging in sight. In this way, programmers repeatedly free themselves from the need to engage in increasingly higher-level programming, thereby enabling them to focus on yet *higher*-level programming. Fundamentally, programming a computer to do something useful today requires setting physical switches in the computer's memory just as much as it did fifty years ago. What has changed is the extent to which the switch-setting process is automated. The result of increasing automation is that humans can spend their days engaged in ever-higher-level aspects of the inventive process, leaving the drudgery to the machine.[33] Although Edison may have been right that genius was 1 percent inspiration and 99 percent perspiration in *his* day, computer scientists have been doing everything in their power to shift more and more of the perspiration from human to machine.[34]

4 Artificial Invention Today

THANK A COMPUTER the next time you brush your teeth to pearly white-ness—at least if you use the Oral-B CrossAction toothbrush, because the crossed bristles at its heart were designed by software called the Creativity Machine.[1] When Gillette, the parent company of Oral-B, contacted Stephen Thaler at Imagination Engines to design its next-generation toothbrush, Dr. Thaler didn't retreat to his workshop to build prototypes using plastic molds and nylon fibers. After all, he isn't a toothbrush designer but a physicist by training.

Instead, he created digital models of a set of existing toothbrushes and gave those models to the Creativity Machine. The particular models he used are called "parametric designs" because they describe each toothbrush using a set of parameters—numerical values representing features of the toothbrush such as the spacing, angle, and stiffness of the bristles, just as a soldier might iden-tify himself by name, rank, and serial number. (The actual models Dr. Thaler used had about 80 such parameters.) By furnishing such information about existing toothbrushes to the Creativity Machine, he did not intentionally bias the Creativity Machine in favor of any particular kind of design; he merely de-scribed the shapes and other characteristics of those existing toothbrushes.[2]

Robots then used the existing toothbrushes to brush fake teeth that had been covered in dye. The results for each toothbrush were reported back to the Creativity Machine in terms of two pieces of information: amount of dye removed and depth of penetration. Once again, neither Dr. Thaler nor anyone else used any knowledge about toothbrush design to steer the Creativity Machine toward or away from a particular design.[3]

Then Dr. Thaler gave the Creativity Machine one last instruction: "Dream."

And so it dreamt. Starting from the existing toothbrush designs and brushing performance data, the Creativity Machine quietly hallucinated, morphing the images of dancing toothbrushes in its digital brain into three-dimensional models of new toothbrushes. Some of these toothbrushes were idle daydreams, but deep within the haze of digital unconsciousness were some dreams worth recalling on waking. To separate the wheat from the chaff, the Creativity Machine autonomously found patterns in the data given to it, essentially learning which design parameter values made some toothbrushes work better than others. The Creativity Machine then used this newfound knowledge to weed out potential designs that were not likely to perform well.[4] Eventually it awoke from the inventive slumber and announced it had dreamt well and produced a toothbrush in which the bristles were crossed against each other, a design that would produce cleaner teeth in a new and unexpected way. So emerged the core of the CrossAction toothbrush.[5]

The Creativity Machine is no one-trick pony. It has been used to write music and literature as well as invent chemical compounds, artificial intelligence for battlefield robots, and warheads for clients including the U.S. Air Force. Despite these successes, Dr. Thaler told me that one of his biggest marketing problems is convincing his potential clients that software can actually solve problems that have eluded the best human minds.

What the skeptics don't realize is that Dr. Thaler designed the Creativity Machine to *mimic* how human minds engage in creative thought, using artificial "neural networks" that are intended to model the interconnections and interactions among neurons in the human brain.[6] Computer scientists have used artificial neural networks for decades to tackle problems where "traditional" computer programming has failed. Traditional computer programming requires the human programmer to program the computer with the specific steps necessary to solve a problem. This works well for writing software that calculates square roots, for example, because the precise sequence of steps needed to do so can be written down in a language that both humans and computers can understand. This kind of programming, however, breaks down for problems such as distinguishing male from female faces, which humans solve by intuition rather than logic. We are unable to describe such solutions as a precise sequence of steps. If the answer to the question, "How do you recognize X?" is "I know it when I see it!" then neural networks are a good candidate for enabling a computer to recognize X.[7]

To use an artificial neural network to recognize a particular pattern, you first "train" the neural network to identify the pattern in much the same way you would train a human: by giving it examples and telling it which ones fit the pattern and which don't. For example, you would train neural network software to distinguish male from female faces by inputting digital images of faces and telling the software which faces are male and which are female. In response to this gentle prodding, the neural network would gradually glean patterns from the examples it observed, such as the higher cheekbones in female faces, without the need for a human to point out those patterns to the software explicitly. The human trainer might not even know what those patterns are. Just as an attentive human trained in this way would eventually develop the ability to distinguish male from female faces, training the artificial neural network would forge connections among the network's virtual neurons that reflect the features distinguishing male from female faces, even though no one had ever directly told the software what those features are.

The neural network is now ready to graduate from school. Once you have trained it, you can show it a face it has never seen before, and it can tell you whether the face is male or female. It pulls off this trick by letting a digital version of the new image flow through the network, triggering some neurons to fire rather than others according to interactions between the image and the connections that were formed during the training process. The end result is an answer ("male" or "female") that emerges from as little conscious rule following as when we perform the same task on people we pass in the street every day.

The Creativity Machine, however, can not only recognize existing patterns but create new ones by using *two* special neural networks—what Dr. Thaler calls "the Imagination Engine" and "the Perceptron." To create the design for the CrossAction toothbrush, he first trained the Imagination Engine by feeding it existing toothbrush designs. Next, he set the Imagination Engine loose, causing it to introduce semirandom modifications into its own neural connections in a way that is intended to reflect the processes occurring in the human brain when it is engaged in creative activity. Perturbing the connections in the Imagination Engine caused it to produce modified versions of the original toothbrush designs that Dr. Thaler had supplied.

Some of these designs represented better toothbrushes than others. This is where the second neural network, the Perceptron, came into play. The Perceptron used the brushing performance data obtained from the tooth-brushing robots to "learn" what makes some toothbrushes better than others. It then used this newfound knowledge to pick out particularly good toothbrush designs from the possibilities produced by the Imagination Engine. The Creativity Machine repeated this process of generating and evaluating potential designs until Dr. Thaler was satisfied with the results.

In Figure 5, which shows the Creativity Machine producing a design for the CrossAction toothbrush, we can see the genie at play. Dr. Thaler didn't design the cross-action configuration of the bristles in anything resembling the way an engineer of yore would have done. He didn't sketch different designs, some with crossed bristles and others without. He didn't build prototypes. He didn't do anything that we would normally think of as designing a toothbrush. Instead, he merely gathered existing toothbrushes, tested them to compile objective brushing performance data, and gave both the toothbrush designs and performance data to the Creativity Machine. Probably the best evidence that he didn't design the physical structure of the CrossAction toothbrush is that when the Creativity Machine presented its creation, he himself was surprised by the results.[8]

In other words, Dr. Thaler invented the cross-bristle configuration of the CrossAction toothbrush design by making a wish for a better toothbrush—in the form of existing toothbrush designs and brushing performance data—to the Creativity Machine. The resulting end product, the wish come true, was the crossed-bristle toothbrush design that the Creativity Machine gen-

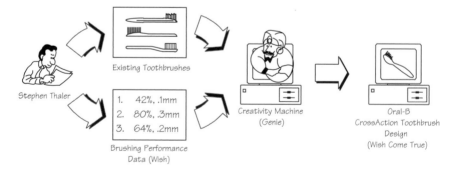

Existing Toothbrushes

Stephen Thaler

1. 42%, .1mm
2. 80%, .3mm
3. 64%, .2mm

Brushing Performance
Data (Wish)

Creativity Machine
(Genie)

Oral-B
CrossAction Toothbrush
Design
(Wish Come True)

FIGURE 5 The Creativity Machine grants a wish. Courtesy Spinney Associates

erated. This and many of the other examples of artificial wishes that we will encounter differ from some wishes that you might make to a mythical genie. A wish for a 1966 Chevy Nova leaves no doubt about the precise machine you want the genie to materialize for you. When you make such a wish, you look to the genie merely as a kind of automated *manufacturing* facility, since you do not call on the genie to *design* anything. You might even refer to the car that the genie produces for you as your "wish."

In contrast, Dr. Thaler's wish left unspecified certain critical components of the toothbrush for which he was wishing, thereby leaving such details for the genie to fill in. With such *abstract wishes*, there is a clear distinction between the *wish* (the instructions that the human hands to the genie) and the *wish come true* (the design or physical object the genie produces when it grants the wish). Abstract wishes, rather than their detailed cousins, are the focus of the remainder of this book.

Invention by Artificial Selection

John Koza is most well known as a computer scientist, the developer of the computer programming technique known as "genetic programming," and a proponent of genetic programming as an automated invention machine.[9] Fewer know him as the inventor of the scratch lottery ticket in the early 1970s.

Genetic programming is a technique that enables a computer to write computer programs *automatically*. As shown in Figure 6, once you create software that can perform genetic programming, you can give the software a high-level description of the *problem you want to solve*, in response to which the software will, if the conditions are right, write a program that solves the

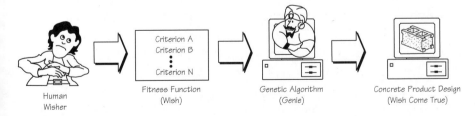

Human Wisher · Fitness Function (Wish) · Genetic Algorithm (Genie) · Concrete Product Design (Wish Come True)

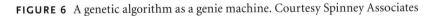

FIGURE 6 A genetic algorithm as a genie machine. Courtesy Spinney Associates

problem.[10] The genie rears its head again. Furthermore, Dr. Koza has devised a variety of techniques for using genetic programming to write programs that specify the physical structure of circuits and other devices. As a result, he has used genetic programming to create not only software but designs for nuts-and-bolts machines as well.[11]

I will focus here on just one of them: the patented controller that we first saw in the Introduction.[12] The term *controller* refers generically to any device that controls another device. The antilock braking system we saw earlier is a kind of controller; it controls the brakes of a car, as is the car's thermostat and its cruise control system. Look inside nearly every mechanical, electrical, hydraulic, or other device in your home or workplace and you will find a controller.[13] Dr. Koza set out to use genetic programming to design an improved version of a particular kind of controller: the *proportional-integral-derivative* (PID) controller, whose name refers to the functions performed by its three primary components, the details of which are not important for our purposes.

The first PID controller design was patented in 1939 by Albert Callender and Allan Stevenson. At the time, the PID controller represented a significant advance over previous controller designs,[14] and PID controllers are still in common use today.[15] Control engineers, however, continue to recognize the need for improved PID controllers.[16]

Dr. Koza and his team succeeded in using genetic programming to automatically produce not just one but several PID controllers that were better than any previous design. One of the controllers they created was 2.42 times better than a controller designed by another team of inventors back in 1998, as measured by those inventors' own criteria.[17]

Genetic programming is one of many kinds of "evolutionary computation," all of which are motivated by the proven ability of biological evolution to develop organisms that are adapted to survive in their natural environment.[18] If natural selection has been able to produce creatures that can run, jump, swim, fly, and survive in just about any environment the natural world can produce, then why can't the basic principles of evolution be harnessed to develop a controller for your car's antilock braking system?

To understand how evolutionary computation might be used to "evolve" a hammer, imagine a world in which hammers appeared spontaneously out of the primordial ooze. Most of the hammers in the first generation, like most of the single-celled organisms that we call our ancestors, would barely be fit for hammering anything. Some would have handles only an inch long; others would lack a head.

But there might be one hammer having a rough handle shape and a tail end that was relatively hard compared to the rest of the body—the makings of a head. In this imaginary world the laws of "nature" are tailored to favor those hammers that are better-suited for driving nails than others. As a result, misshapen hammers tend to die before they can mate, while hammers with features that are useful for driving nails tend to survive. But they also do more. They mate, producing offspring. A child hammer might inherit the handlelike body of its father and the hard head of its mother—and thus be an even better candidate for driving nails than either of its parents individually. Another child, much to its parents' disappointment, has the curlicue body of its mother and the soft head of its father. No one selects it to be on the hammering team at school.

Mom and dad have a third child who doesn't just inherit their characteristics. Its head is even harder than either of its parents—a mutation. This child goes on to mate with another hammer from another family, producing yet more hard-headed offspring.

The circle of life continues. Individuals mate with each other, thereby combining their genetic material to produce offspring, some of whom include further variations as the result of mutation. Because the laws of nature in this hammer-eat-hammer world favor skill at driving nails, successive generations of hammers tend to develop physical features that are better for performing that function.

What I've just described is essentially how genetic programming, and "genetic algorithms" more generally, work.[19] The key to using a genetic algorithm to solve a particular problem is to give the algorithm a set of artificial "laws of nature" that suit the problem to be solved. If you want to use a genetic algorithm to produce a better hammer, then the two key initial conditions you must specify are (1) the "genotype" of a hammer, which represents the features that may vary from one hammer "organism" to another; and (2) the "fitness function," which specifies the criteria for determining whether individual hammers should survive until the next generation. Once you have specified these and some other initial conditions, you can tell the genetic algorithm to start running and it will simulate the evolution of hammers in the manner just described. The key difference between natural evolution and artificial evolution is that the actual laws of nature dictate natural evolution, while a human provides artificial laws of nature that govern artificial evolution.[20]

This is precisely what Dr. Koza and his team did to evolve the improved PID controller mentioned earlier. They set up the artificial world of the genetic programming software with knowledge of the basic components that *any* controller might contain, analogous to stocking the Earth with carbon, hydrogen, oxygen, and the other building blocks of life. Then they defined the fitness function, which specified the *criteria* or *requirements* that a controller would need to satisfy to solve the problem they were trying to tackle. For example, the fitness function described the controller's "transfer function," namely the mathematical relationship between its inputs and output. (The transfer function of an amplifier that takes an input I and produces an output O that is double the value of the input would be $O = 2I$.) The fitness function also specified the maximum amount of "overshoot" to be produced by the controller under certain conditions. A thermostat without such limits on overshoot might first send your home's temperature skyrocketing from 60 degrees to 120 degrees before settling down at the more comfortable 70 degrees you specified.[21]

Critically, however, the fitness function did *not* specify such important details as *which particular components* to include in the controller, how to connect those components to each other, or how to fine-tune those components to solve the problem at hand. The fitness function did not, in other words, specify the physical design of any particular controller. The fitness function did not even specify that a PID controller, rather than some other kind of controller, should be used to solve the problem. Instead, Dr. Koza "started with the high-level design requirements of the problem and gave genetic programming a free hand to create any topology that satisfies the problem's requirements."[22]

Before running the genetic programming software, therefore, the fitness function merely represented Dr. Koza's *wish* for a controller that would satisfy the criteria laid out in the fitness function. Now that you have seen some examples of the criteria that were included in the fitness function, you can also see how the genetic programming software could use those criteria to evolve controllers. In creating an initial random population of controllers, it used a *simulator* to run a battery of predetermined virtual tests on those controllers. From the results of the simulations for a particular controller, the software could determine how well the controller scored according to each criterion specified by the fitness function. For example, the software could determine whether the controller overshot its mark by more than the 2 percent limit specified by the fitness function when the simulated controller was given a particular kind of signal. Controllers staying within the overshoot limit were given lower scores on the overshoot criterion than those that didn't. The software combined all of the scores for a particular controller into an overall grade for the controller. Then the software compared grades among controllers to determine which of them would be more likely to mate and survive until the next generation.

The feature of the fitness function that enables it to produce a solution to a problem, without defining ahead of time *how* to solve the problem or *what* the specific solution should be, is that the fitness function specifies criteria, sometimes called requirements, that can be used to evaluate potential solutions to the problem, and thereby to compare the fitness of one potential solution in relation to another. To find the best light bulb filament, Thomas Edison didn't need to know ahead of time which filaments would work best. He didn't even need to understand anything about the physical mechanisms by which sending an electrical signal through a filament produced light. All he needed to know was how to *test* individual filaments—namely by sending electrical signals through them—and *evaluate* the results by observing and determining which filaments burned brighter and longer than others.

Dr. Koza's simulated evolution produced a controller that did not use the set of components used by conventional PID controllers. Instead, the controller that emerged at the top of the artificial food chain used two "derivative" blocks, whereas conventional PID controllers use only one.[23] This is an example of how artificial invention technology, unconstrained by the preconceived notions or biases of human inventors, can produce designs that fall outside the box of conventional human design principles.

Genetic algorithms and other forms of evolutionary computation are probably the most widely used kind of artificial invention technology today. For example, the NASA antenna that we encountered earlier was evolved using an evolutionary algorithm.[24] The fitness function for the antenna specified, for example, that the antenna should have a voltage standing wave ratio—the ratio between the maximum and minimum voltage in a standing wave pattern—of less than 1.2 while transmitting signals and less than 1.5 while receiving them.[25] Again, such a fitness function is a kind of artificial wish, as I use the term, because it specified criteria the antenna should satisfy, without specifying the physical structure that the antenna should use to satisfy those criteria.

Here are just a few more of the tastiest examples of what evolutionary computation has been used to invent to date:

Natural Selection developed an evolutionary algorithm for Agouron Pharmaceuticals (subsequently acquired by Pfizer) to find the optimum confirmation (shape and position) for a candidate drug to bind into a target protein binding site. The software produced such a significant improvement in the efficiency of the drug discovery process that it has become the most-used software by 1,500 computational chemists at Pfizer.[26]

Matrix Advanced Solutions used its proprietary software to develop an oral thrombin inhibitor to act as an anticoagulant. The thrombin inhibitor, which was developed without the use of any expert knowledge, is now in preclinical trials.[27]

Icosystem's "Hunch Engine" is being used to help discover molecules for use in new drugs by weeding out those that are unlikely to be useful before presenting the best candidates to a chemist for evaluation. Once the chemist selects molecules that appear promising, the Hunch Engine uses those selections to generate a new set of candidate molecules for evaluation.[28]

Hitachi used genetic algorithms to evolve a nosecone design having optimal aerodynamic performance for the new Japanese bullet train (*Shinkansen*), thereby minimizing the sound inside and outside of the train, while holding the maximum number of passengers.[29]

Peter Senecal at the University of Wisconsin-Madison used genetic algorithms to design a new piston geometry to reduce the fuel consumption of diesel engines by 15 percent, reduce nitric oxide emissions threefold, and reduce soot emissions by 50 percent over the best available technology.[30]

Maxygen Biotech used genetic algorithms to design a Hepatitis C treatment that entered clinical trials—the first such protein to enter clinical development.[31]

Although genetic algorithms have been around since at least the early 1960s,[32] it is only recently that they have evolved (pun intended) the ability to produce real-world products reliably across a wide range of problems. Although hard data are difficult to obtain, a recent informal survey indicated that the number of people entering the field of evolutionary computation has grown exponentially since the 1960s.[33] Many people, however, including many computer scientists, still think genetic algorithms and other forms of artificial invention technology are not ready for prime time. As the examples given here illustrate, nothing could be further from the truth.

5 Invention Automation

Old and New

EVEN TODAY'S MOST POWERFUL invention automation technology auto-mates only *part* of the inventive process. Genetic programming software, which writes the very source code that previously only human programmers were capable of writing,[1] requires a human to write the fitness function and set up the other initial conditions that the genetic programming software needs if it is to do its thing.

To understand in more detail how different kinds of artificial invention technology automate parts of the inventive process, we need some way of breaking down the process into components, so that we can identify which one has been automated by a particular technology. For this purpose, let me introduce a common way of thinking about the design process, and problem solving more generally, called the "Waterfall Model," as shown in Figure 7.[2]

The sequence of six steps illustrated in Figure 7 can be followed from top to bottom to design a toothbrush or to solve other kinds of problems. The key feature of this model is that you use it by first formulating an abstract defini-tion of the problem to be solved and then proceeding to specify the solution in increasingly specific, detailed terms until the end product (such as a new kind of toothbrush) is built and put into use in brushing teeth.

Consider application of the waterfall model to invent a better mousetrap. You would begin with the Problem Definition phase by defining the problem as "to design a mousetrap that catches mice more effectively." Next, in the Require-ments phase, you might specify that the mousetrap must (1) lure mice within a

20-foot radius of the trap, (2) catch mice without killing or injuring them, and (3) make it impossible for mice to escape.[3] Note that this set of requirements does not yet specify any physical details of the mousetrap you are designing.

Next comes the Functional Design stage, where you specify what the mousetrap must *do* to satisfy the requirements you have just laid out.[4] For example, you might indicate that the mousetrap must perform the functions

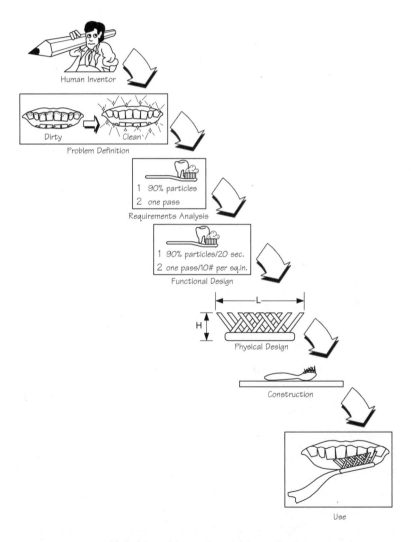

FIGURE 7 Old-fashioned inventing. Courtesy Spinney Associates

of (1) storing food that is attractive to mice, (2) releasing the odor of the food within at least a 20-foot radius of the trap, (3) ensnaring the mouse when it comes into contact with the food, and (4) affording a sufficient air supply to the mouse while it is trapped.

Next comes the Physical Design (sometimes called Structural Design) stage, where you specify particular physical components for solving the specified problem, satisfying the specified requirements, and performing the specified functions. For example, you might draw a diagram of a contraption having a cheese holder, a small fan to spread the odor of the cheese, a trap door triggered to close by the weight of the mouse, and vent holes for allowing the mouse to breathe while inside the trap. In the Construction phase, you build the mousetrap you've just designed, and in the Use phase you lay it on the kitchen floor.

I've labeled Figure 7 "old-fashioned inventing" because traditionally a human inventor (or team of inventors) did all of the legwork up to, and possibly including, Construction. Even if the inventor did not ploddingly engage in separate stages of Problem Definition, Requirements Analysis, and Functional Design in the textbook manner of Figure 7, he or she at least performed the critical step of Physical Design. The act of designing[5] a particular physical structure to solve a problem is the essence of what we have meant historically by "inventing." Traditionally, we call those who excel at Physical Design "inventors," those who can only build products designed by others "mechanics" or "builders," and those who merely engage in the more abstract stages of Problem Definition, Requirements Analysis, and Functional Design "philosophers," "idea people," or (worse yet) "managers."

Invention Automation Swims up the Waterfall

This progression from abstract to concrete, from general to specific, illustrates precisely how artificial invention technology is automating invention, even when such automation is only partial. Although John Koza and his team still engaged in significant work to create their controller—they wrote a fitness function and set up the initial conditions for artificial evolution to run its course—there is one thing they decidedly did *not* do: design the physical structure of the controller. It was the genetic programming software that performed this crucial step, as shown in Figure 8, in which human-performed actions are on the left and computer-performed actions are on the right. Once

Dr. Koza and his team had defined the problem and analyzed the requirements for a better controller, they handed off the baton to the genetic programming software, which performed the essential step of Physical Design
automatically, by evolving a design for a particular controller that satisfied the
criteria laid out by Dr. Koza and his team.

To understand the significance of such automation of physical design, take
yourself back to the Stone Age. To create a new arrowhead or other physical
device, you had to manually construct it yourself. In fact, you could view inventing in the Stone Age as a problem in physical construction.

Now fast-forward to the Industrial Age, in which technologies including
the assembly line automated the process of Construction, the penultimate tier
of the waterfall. As a result of widespread adoption of such automated manufacturing techniques, inventors no longer needed to be builders. Instead, an
inventor could invent a machine simply by designing its physical structure—

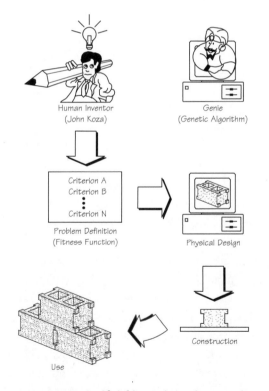

FIGURE 8 Artificial invention using genetic
algorithms. Courtesy Spinney Associates

the next-highest tier in the waterfall—and then hand off the design, in the form of verbal or written instructions called a *structural specification*, to a builder or a manufacturing company, who could then build the machine automatically (or at least predictably) without any additional inventive effort.

Automating one tier in the waterfall frees inventors from the need to perform the work of that tier. As a result, once the work of a new, higher tier of the waterfall has been automated, the inventor's job *ends* one tier higher than it did previously. A bedrock principle of patent law is that the "true date of invention is at the point where the work of the inventor ceases and the work of the mechanic begins."[6] Improvements in invention automation technology have been moving that point of division between the inventor and the mechanic further and further up the waterfall.

The introduction and widespread adoption of general-purpose computers—universal machines—did the same trick for Physical Design as Industrial Age technology did for Construction. With computers at their disposal, programmers no longer needed to engage in the Physical Design of machines to turn their ideas into reality. Instead, they could stop working and go home after completing the higher-level tier of Functional Design,[7] since writing a computer program is an exercise in functional design—computer programs describe the *functions* a computer is to perform, but not the *physical mechanisms* by which the computer is to perform them. Computers shifted the dividing line between the inventor and the mechanic one level higher in the waterfall.

The latest generation of artificial invention technology takes this progression to its next logical stage, pushing the dividing line yet higher up in the waterfall. John Koza used genetic programming software to design his new controller merely by engaging in Problem Definition[8] and then furnishing the results of that effort, in the form of a fitness function, to the software. He did not need to engage in the functional design that would have been required by older styles of programming. The best evidence of the fact that genetic programming automates Functional Design is that it produces computer programs, and that writing such programs is a kind of functional design.

Breaking down the inventive process into stages, and identifying the stage(s) that have been automated, makes clear that even *partial* automation of the inventive process can have far-reaching effects on the process of inventing. Automating even physical design means that human inventors no longer need to perform a task that traditionally has defined what it means to be an inventor. Despite the common focus on when or whether machines will be able to *completely* replace human inventors,[9] we don't need to wait for the arrival of a full-blown artificial intelligence to experience computer automation of invention; it's already been here for decades.

Two Types of Automation

Existing invention automation technology employs two basic kinds of techniques for automatically creating designs, which correspond more generally to two ways of solving problems. The first is problem solving by *design*, which involves rigidly following a set of rules to arrive methodically at a solution to the problem. The waterfall process is a quintessential example of problem solving by design. Examples are assemblers, compilers, synthesizers, and other invention automation technologies that replace high-level statements with predetermined sets of low-level details. The defining feature of design-based automation is that it is *deterministic*; once a description has been written in the right language, a machine automatically translates that description into a final design using rules that leave little or no room for interpretation, just as you follow the cookbook instruction to "boil water *as described on page 4*" by going to page 4 and following each and every instruction you find there.

The other basic problem-solving technique is *search*,[10] which involves generating possible solutions to the problem (perhaps at random), testing them all, weeding out the ones that don't solve the problem, modifying or adding to the remaining potential solutions, and repeating until a satisfactory solution is found. Randomness and brute force trial and error, rather than the stern guiding hand of an all-knowing designer, play the primary role here. Genetic algorithms and the Creativity Machine are exemplars of this approach. Thomas Edison was a master of search. He found his first light bulb filament by searching for it—he hired experts to scour the globe for source materials, testing more than 6,000 of them before settling on carbonized bamboo. (Tungsten didn't come until later.)

The dividing line between design and search is far from sharp. In fact, each strategy necessarily includes elements of the other; they are invention's yin and yang.[11] For example, although I described the waterfall process as if it leaves nothing to chance, this is overly simplified. At each tier of the waterfall, you have a *choice* about how to carry out the task specified by that tier. No rigid rule required John Koza to specify a 2 percent overshoot as part of the requirements analysis for his controller. Computer programmers can choose which instructions to include in a sorting program. Automotive engineers can choose which kinds of metal to use for a car frame. At each stage, inventors inevitably pick some initial possibilities, evaluate them, and then modify the initial choices and repeat. Although some rigid rules do govern the waterfall process—such as the rule that each stage must be completed before the next begins—*within* these rigid rules lies room for a healthy dose of *searching* for elements consistent with those rules.

Similarly, even the most randomness-infused search-based process is governed by rules. Natural evolution is governed by the laws of nature. Artificial evolution, as carried out by genetic algorithms, is governed by the artificial laws specified by human programmers. Rigid structure constitutes a framework within which a search operates. Without such a structure, the search would not be *for* anything and would never arrive at a result.

Furthermore, any computer-based search process that is used to solve a complex problem must have a heavy dose of design baked into it if it is to complete its task within the lifetime of the universe; it is not feasible to search through *every* possible solution to real-world problems. Exploring every possible path even through the relatively simple game of checkers would require considering about 10^{40} possible moves. Exploring three moves per millisecond would therefore require 1,000,000,000,000,000,000,000 centuries.[12] Today's artificial invention researchers therefore spend much of their time refining genetic algorithms and other search-based techniques to search through the "space" of possible solutions as efficiently as possible. One avenue that has borne much fruit has been to tailor genetic algorithms to construct solutions from basic building blocks, which are interconnected in modular hierarchies of increasing abstraction and power.[13] This should sound familiar; it is how programmers write increasingly powerful programs. Furthermore, it is how natural evolution has been able to evolve the human eye and other biological structures that never could have evolved through a purely random process of trial and error at the molecular level, even on geologic time scales.[14]

Implications of Automation

Although design-based automation and search-based automation differ in *how* they automate inventing, they have many of the same effects. If you are an inventor with either kind of automation at your disposal, then from your point of view the task of inventing has been reduced to the job of performing the highest tier in the waterfall directly *above* the one that has been automated, whether the automation is performed deterministically or by search.

If It's Cheaper I'll Try More

Providing an inventor with automation technology that can not only fill in low-level details of a design but do so *quickly* enables the inventor to generate and experiment with a larger number and wider variety of designs in a given period of time. If the inventor can generate a prototype every two hours instead of every eight hours, she can now test four prototypes a day instead of one. As a result, increasing the *speed* with which inventors can produce potential solutions to problems enables them to solve problems more quickly and also produces *better* solutions in the same time than was possible previously. In other words, sometimes a quantitative increase in speed can bring about a qualitative improvement in results.

Increases in computing speed had exactly this effect on computer programming. If you were a programmer in the early 1960s, you would encode your program on boxes full of punchcards and then drop the cards off at a data processing center housing a single powerful computer. Sometime later—maybe a day, maybe a week—a computer operator would run your program and produce a printout of the results, which you would pick up and review. If you then discovered a bug in your program, you would have to produce a new set of punchcards and repeat the entire process.[15]

The glacial pace of such "batch processing" encouraged programmers to write programs conservatively so as to avoid introducing any bugs. Furthermore, the need to wait so long between program iterations meant that programmers could make only so many improvements to their programs before completing the project on schedule. This limited both the size and scope of programs that could be produced.[16]

In the 1960s, more powerful computers were developed that allowed programmers to write a program and then run it and obtain the results immediately—no trip to the data processing center required. Such technology didn't just enable programmers to write the *same* programs more quickly. Rather, programmers began to change *how* they wrote programs, adopting a more interactive and experimental programming style. They started writing just *portions* of programs to test and refine them, thereby allowing them to catch and fix bugs sooner and more easily. More generally, "by eliminating the 'wait

and see' aspect of batch-processing, time-sharing made it possible for programmers to treat their craft as a performing art."[17] Once keyboard-based
input became available, programmers who had lived through the batch-
processing days viewed "the ability to type a command on a keyboard and see
the computer's immediate response on their own printer [as] nothing short
of miraculous."[18]

Giving programmers technology for instantly and automatically transforming their written code into running software enables them to use the
computer as sculptors use clay, allowing them to sustain the state of unin-
terrupted flow that is so essential to creative work.[19] Therefore, although increased computer interactivity does not automate programming *per se*, shifting the burden of filling in low-level details from the human programmer to
the computer enables the programmer to focus continuously on the creative
aspects of programming rather than on the administrative work that filled
J.C.R. Licklider's early days. This has led in recent years to the development of
"extreme programming" and "agile programming" techniques, both of which
place a high value on quickly developing interim versions of software, revising
it on the basis of feedback obtained from other programmers or customers,
and repeating as necessary.[20]

You're familiar with this ability of increased speed and interactivity to enhance creativity just from writing documents using a word processor rather
than a typewriter. Most of us readily use word processors to pound out long
drafts containing our initial, rough thoughts. Then we review those drafts
and revise them directly in the electronic document itself. We don't have the
same need to plan every word as carefully as we did in the days of the typewriter; we can quickly and cheaply fix any errors or omissions after the fact.

Although these features of word processors can degrade our writing skills,
they also can positively influence our creativity. The power to produce five
alternative drafts of an essay in the time previously required to produce just
one enables you to explore a wider variety of themes for the essay. Because you
can modify your thoughts on the screen essentially instantaneously, you can
use the words in front of you as a virtual laboratory, testing and evaluating
pathways that may not have occurred to you if you were limited to generating
them in your mind alone.

Electrical engineers already have begun to realize the same benefits from
the electronic design automation (EDA) technology that enables them to create circuit designs merely by writing instructions in hardware description

languages (HDLs). Those engineers can try out alternative designs quickly simply by tweaking the HDL code, compiling it, and using a simulator to perform virtual tests of the resulting circuitry.[21] They can modify a circuit design not by adding or moving around capacitors or resistors on a breadboard or in a circuit diagram but by editing the HDL code. They can even incorporate entire modules from previous designs just by cutting and pasting HDL code, in the way that you cut and paste text from one report into another. These benefits of inventing by programming are so significant that we should expect to see it adopted in other fields facing complex design problems. For example, modified versions of HDLs may be used to develop tomorrow's nanoscale devices, which no human could hope to design molecule by molecule.[22]

The process of inventing by programming, coupled with high-speed computer automation technology that makes it possible to iterate through multiple design possibilities quickly, can do more than speed up the development process; it can improve the quality of the results obtained. For example, Jeff Walker of TenFold related to me how the company's software, in addition to enabling businesspeople to create software without knowing how to program, also yields "an additional benefit that we [at TenFold] call 'power.'"

> "Power" refers to the functional richness of the application, in other words, what it does. Say that your application lets someone use it over the Internet, or your application automatically sends email to someone when you want it to. . . . I could go on with thousands and thousands of features—that's what I call power. If you build an application with TenFold's technology, the power that you obtain with that application, even though you're building it ten times faster, is dramatically better. The benefits of the automation are a much shorter [development] time period, much lower cost, and . . . much higher-quality and much more powerful products.[23]

Mind Extension

Shifting responsibility to invention automation technology for handling low-level details can enable inventors to create higher-quality inventions than they could have produced otherwise. One reason is that human minds can handle only so much detail before becoming overloaded. In *theory*, given unlimited time, electrical engineers could design today's billion-transistor microprocessors manually. They could manage the complexity of the design process by importing techniques used by authors to write multivolume treatises into the microprocessor design process: writing a high-level outline, breaking it into sections, and dividing responsibility for the parts of the circuit among engineers. The waterfall model is one example of this process, which is sometimes called *reduction, decomposition,* or *analysis.* The process of combining the resulting pieces into an integrated solution is referred to as *synthesis.*

The technique of "structured programming," introduced in the 1960s precisely to encourage programs to be written in this way, survives to this day.[24] Similarly, John F. Jacobs at the Lincoln Laboratory introduced engineering-style top-down programming techniques into the development of the massive SAGE air defense system in 1955. As historian Martin Campbell-Kelly observes, "The exodus of programmers from the SAGE project in the 1950s caused this project-management style to diffuse throughout the software industry" for decades.[25]

Such manual techniques, however, have practical limits. They can be used to write a 300-page book, or even a 26-volume encyclopedia, but not to design a billion-transistor microprocessor. The sheer complexity of the problem mandates that human designers use technology to *automate* the low-level details so that they can focus their (limited) minds on a subset of the outline— its highest levels, which is a much more manageable problem. Genetic algorithms already have proven they can be used to apply search-based strategies to such a task by solving a problem with a billion variables.[26]

Inventors stand to benefit doubly from such technology because it allows them to produce *more* powerful inventions with *less* total human effort. There's no contradiction here. As computer scientist Edsger W. Dijkstra said:

> It has been suggested that there is some kind of law of nature telling us that the amount of intellectual effort needed grows with the square of program length. But, thank goodness, no one has been able to prove this law. And this is because it need not be true. We all know that the only mental tool by means of which a very finite piece of reasoning can cover a myriad cases is called "abstraction"; as a result the effective exploitation of his powers of abstraction must be regarded as one of the most vital activities of a competent programmer.[27]

Everywhere, But Invisible

Decreases in the size and cost of computers further reinforce the impact of increases in speed and interactivity on the process of inventing.[28] For example, as computers became smaller and less expensive, it became feasible for every programmer to have his own computer at which to program full-time—a much more interactive experience than sharing a single computer with other programmers.

However, making computers cheaper, smaller, faster, and more ubiquitous can have another, somewhat paradoxical, effect. As computers increasingly pervade every physical space we occupy and every task we perform, you might think they would increasingly occupy our every conscious thought. Yet as they become smaller, faster, less expensive, more interactive, and more ubiquitous, they can *fade* from our consciousness, and even become all but invisible.

You are familiar with this phenomenon just from driving a car. The first time you sat in the driver's seat, you focused all your mental energy on the steering wheel, gas pedal, brake pedal, and other physical components that you needed to control. The driving instructor probably even had to remind you to "keep your eyes on the road" and not on the wheel. As you became more proficient at driving, however, you needed to pay less and less attention to your physical actions. You gained the ability to translate your desires—turn left, slow down, get through that yellow light before it turns red—into the necessary physical actions automatically and without conscious thought. Now you can even drive, talk on your cell phone, and eat a slice of pizza all at the same time, without thinking about the steering wheel or the gas pedal, which have become invisible to you.

Computers have become increasingly invisible to programmers and users for the same reasons. We already saw, in connection with the Mentalist Fallacy, that the tasks programmers perform have become so far removed from the internal physical operation of computers that programmers can sometimes forget—or never learn in the first place—that the software they write has any physical form at all. Programmers refer to software as "intangible" or as a kind of "pure thought." This is the ultimate in invisibility, and it is what enables programmers to devote 100 percent of their mental energies to what they do best—designing and writing better programs—rather than to the physical mechanisms for creating and executing those programs.

Such mental invisibility, in which you *physically* see the tool you are using but in which the tool does not occupy your conscious thought, is being reinforced by the increasing *physical* invisibility of computers as they shrink in size. A computer hidden underneath your desk or tucked away in your shirt pocket can remain *mentally* invisible much more easily than a refrigerator-sized computer, no matter how seamlessly you may interact with the latter.

Computer invisibility is no accident; it results from the conscious efforts of generations of computer professionals, culminating in a movement now known alternatively as "invisible computing," "pervasive computing," and "ubiquitous computing."[29] The invisible computing movement seeks to create a world in which computers are *everywhere*—embodied in radio identification tags in every manufactured object, in every wall of every building, even embedded in our bodies. Computers, according to this vision, will become so fully integrated into our experience that they are at once all-encompassing and completely invisible.[30] As Mark Weiser, the intellectual progenitor of pervasive computing, noted:

> A good tool is an invisible tool. By invisible, I mean that the tool does not intrude on your consciousness; you focus on the task, not the tool. Eyeglasses are a good tool—you look at the world, not the eyeglasses.[31]

Increasingly invisible computer technology can potentially enable integration of computers into the inventive process more seamlessly than ever before. "There will come a time, perhaps, when we take the power of computational innovation for granted, as we now barely notice the power of mechanical or electrical machines."[32] If implemented appropriately, such technology promises to make inventing a more human activity, even as the inventive process becomes increasingly imbued with the power of machines.

One Wish, Many Inventions

Artificial invention technology promises to reduce the workload of inventors not just at the time they design a particular toothbrush or car frame. If an inventor writes a solid wish, the wish will pay dividends far into the future. Imagine you are John Koza in 1999, having just constructed a 1,000-Pentium supercomputer for $450,000. You want to design a new controller, so you write a wish for a controller in the form of a fitness function. You hand that wish to an artificial invention genie running on your supercomputer and instruct it to "Invent!" You have a one-month deadline to supply the controller design to a customer, so you go away on vacation and return exactly one month later and tell the genie to stop working and give you the best controller design it has found so far. The design you obtain in this way is good enough for your customer, but you suspect that an even better one could be found, if only you had more time to let the genie grind away.

Ten years later, you fork over another $450,000 to buy a supercomputer that is 8,000 times more powerful than the Edsel you had in 1999.[33] You have another month on your hands, so you feed in *the same wish you used in 1999* and tell *the same artificial invention software you used in 1999* to grant the wish. The only difference now is that your genie is running on a more powerful computer. You go away on vacation for another month and come back to find that your genie has now produced a much better controller than it did a decade ago.

Consider what has just happened. You wrote your wish *once* and then performed *no additional work*, yet you were able to produce a better controller than before—possibly one that is different enough that we would call it a new invention—simply by using more advanced computer technology to grant your wish. Furthermore, there's nothing to stop you from doing this over and over again as computer technology improves, thereby compounding your returns. Although in this example you used a faster computer to achieve better results, you could also improve the effects of your wish by using a better simulator, an improved genetic algorithm, or some other improved technology, all of which you may have obtained or purchased from someone else, again with no additional inventive effort on your part.

This example illustrates how search-based automation differs qualitatively from most traditional software algorithms. Running most old software on a newer, faster computer will cause the software to run *more quickly*, but not to produce a qualitatively different result from what it did before. Perhaps you've had this experience if you've installed an old videogame on a souped-up new computer, only to find the characters in the videogame racing around the screen at breakneck speed. No matter how much you speed up your computer, however, the videogame remains the same old videogame, merely faster. It will never morph into tax software (thankfully).

Handing a wish to a search-based invention genie doesn't work this way. If you grant such a wish on more powerful hardware, or by using a more powerful genie, you may very well be able to produce better results than you did before, or even to *solve problems that previously you could not solve* in the same amount of time. At the risk of being morbid, your original wish may continue to bear fruit long after you've departed this world.

This is just a taste of why (as we will explore in more detail in Chapter 8) patents on wishes themselves will become particularly valuable in the Artificial Invention Age, why such patents will be tricky to interpret, and why we need to begin now to address any problems in patent law proactively so that we can avoid, or at least minimize, any problems that result from "wish patents."

Design-based automation can beget multiple inventions from a single wish in a different way from search-based automation. Write a program in the C programming language and (at least in theory) you can compile that C source code with a Windows-based compiler to produce software that will run on a Windows PC, and then compile the same source code again with a Mac-based compiler to produce software that will run on a Mac. As a result, you are spared the need to write two versions of the same program. The appeal of this labor-saving ideal was recognized by Sun Microsystems when it released the Java programming language under the slogan, "write once, run anywhere," reflecting the ability to write one Java program that could be compiled to run on *any* platform—Windows, Mac, or Unix.

This ability to use a single computer program to automatically produce software having many forms is known as *multiple realizability*. This simply means a single computer program can be physically embodied or *realized* in multiple pieces of software that differ from each other but nonetheless all perform the same process described by the program.[34]

The von Neumann computer architecture also exhibits multiple realizability. Buy a version of Quicken financial software for Windows and you can install and run it on any Windows-based computer, whether the machine was manufactured by Dell, Sony, or HP. Although the specific hardware on these computers may vary widely from each other in many respects—such as the kind and amount of memory they contain—all of them adhere to the "Wintel" platform (a contraction of Windows and Intel), thereby enabling all of them to run the "same" software.

If software were not multiply realizable in both these ways, mass markets in computer hardware and software might never have developed, or we might have been stuck with one or two very specific models of computer, rather than the great variety we see today. If software developers had to rewrite their software every time Dell released a new computer model, software development would grind to a halt. Honeywell may have been the first company to use multiple realizability as a marketing tool in 1965, when it advertised its "Liberator" software for letting IBM users convert their old IBM 1401 software "and run [it] at high new speeds on a new Honeywell Series 200 computer. Automatically, permanently, painlessly. No reprogramming. No retraining."[35]

Complex Interactions Between Complexity and Effort

One unintuitive consequence of the ability of invention automation technology to shift the burden of designing low-level details from human inventor to computer is that sometimes the technology produces inventions that are more complex and costly to manufacture than an equivalent invention designed directly by a human inventor. Recall our 5-line SquareRoot program, which was translated by an automatic assembler into a 250-line machine language program, which might be incomprehensible to a human programmer. A skilled assembly-language programmer might have been able to write a program to achieve the same result using just 25 easily readable assembly-language instructions.

The high-level programmer, however, probably is willing to accept such a tenfold gain in size and complexity of the *result* in order to obtain a corresponding reduction in the duration and complexity of the programming *process*. Why should the programmer care if the software she creates consists of 250 machine-language instructions that she cannot understand, so long as the software does what she intends it to do, and she is able to create this software more quickly using an automatic assembler than by hand? Microprocessor designers make the same tradeoff when they use EDA software to create designs for processors with billions of transistors. Such a design may be riddled with inefficiencies that could in theory be eliminated by the careful eye of a skilled engineer, but the extra cost of performing such streamlining typically doesn't outweigh the benefits of getting a faster, more powerful, low-cost processor out the door more quickly.

You're familiar with a similar phenomenon even if you're just a user, rather than an inventor, of machines. Modifying a manual-transmission automobile to make it less complex for the *driver to use* involves replacing the car's manual transmission with a much more complex (and expensive) automatic transmission. Increasing the internal complexity of the car in this way is desirable because it reduces the car's complexity *from the point of view of the driver.*

Furthermore, I've been assuming all along that once an artificial genie has produced a *design* for an invention—such as a three-dimensional model of a toothbrush—the invention itself can be manufactured easily or even automatically. This certainly is true for traditional software; once a programmer writes source code, a computer can automatically "manufacture" the corresponding software by flipping the right bits in the computer's memory. Making additional copies of the software is just as effortless thanks to the availability of fast and cheap copying technology and storage media. This is not always the case, however. Artificial invention technology can produce designs that perform their intended function but do so using physical structures more complex and expensive to manufacture than those of their manually designed cousins.

Therefore it is not possible to say in general whether simplifying the design process to make the inventor's life easier will reduce the complexity of the resulting product or the cost of manufacturing it. Different approaches will work best in different circumstances, which is one reason companies in the Artificial Invention Age will need to be familiar with a variety of approaches and mix and match strategies that work best to tackle a particular problem, as we will see in Chapter 12.

Tradeoffs of Search vs. Design

Now that we've seen some of the advantages and disadvantages of inventing-by-wishing in general, it's time to explore some of the features that distinguish design-based strategies from search-based strategies. For example, when you invent by searching you need to know the problem you are trying to solve and how to *recognize* a solution to the problem, but you don't need to know or understand the *form* of the solution, either before or after your search is complete. Edison required only the knowledge that he needed a material that would create light inside a bulb when electricity was pumped through it, and the ability to test each material for that property. He didn't have to know anything about the chemistry or physics of the materials he tested, and even after he found one that worked it wasn't necessary for him to understand *why* it worked, only *that* it did. If you are trying to solve a radically new or complex problem (such as finding a drug that will cure a particular disease even though you don't understand the physical mechanisms by which the disease works), you may not even be able to specify the form or structure of the solution ahead of time. As a result, it may be impossible to write the high-level outline necessary for a design-based approach to work. In this case, a search-based approach may be your best, or only, bet.[36]

Compare this to the much-touted success of IBM's Deep Blue chess-playing machine, which defeated human chess-playing legend Garry Kasparov in 1997. Although this event often is cited as a victory of machine over human,[37] Deep Blue's human designers worked with a human chess grand master to directly program detailed rules for playing expert chess into Deep Blue. The success of Deep Blue, therefore, is attributable to the human chess-playing expertise that its human designers "bottled" inside it. Such a strategy is not applicable to situations in which we do not have such up-front knowledge of how to solve the problem at hand.[38]

This benefit of search-based strategies can also be a drawback. Search-based artificial invention technology often produces results that the inventors themselves *do not understand*. For example, Thinking Machines founder Danny Hillis used evolutionary computation software to create programs for sorting numbers. When you give such a program a list of scrambled numbers, such as 9 8 2 7 3, it gives you back a sorted list of the same numbers: 2 3 7 8 9. Hillis examined the number-sorting programs that his software had evolved but could "not understand how they work. I have carefully examined their instruction sequences, but I do not understand them: I have no simpler explanation of how the programs work than the instruction sequences themselves. It may be that the programs are not understandable."[39] Although such software is useful for sorting numbers, the difficulty (or impossibility) of understanding it comes at a price. For example, the sorting software may have a hidden bug or fail under conditions that we cannot predict since we don't understand how the software works. As a result, you probably wouldn't want to use autopilot software that you don't understand in a space shuttle.[40]

Similarly, a disadvantage of search-based solutions is that they often produce results that include "junk," much as the evolution-derived human body contains an appendix that serves no purpose.[41] As a result, search-based solutions tend to be larger and sloppier than those produced by design-based methods. Moreover, search-based solutions may be one-shot deals in the sense that, because we often cannot gain insights into their workings by studying them, they often afford little basis for making future advances in science or technology.[42] Design-based problem solving, in contrast, tends to lead to solutions that we understand well, precisely because they are designed by proceeding from first principles.

A significant benefit of search-based strategies is that, because they are driven by randomness and empirical evaluations rather than predetermined rules, they are not limited by the standard design principles taught in engineering school or by the blind spots, prejudices, biases, or intuitions of human inventors.[43] As a result, they can produce inventions that human inventors would be unlikely to produce, as evidenced by the fact that human experts are often surprised by those inventions.[44] The team that used evolutionary computation to design the NASA antenna we encountered earlier showed the antenna to trained antenna engineers, who "didn't have the faintest idea of how or why it worked. For the life of them, they'd never think to bend wires in this weird way. It looked to them like someone dropped a paper clip and stepped on it."[45]

The ability of search-based invention automation software to produce results that surprise even human experts demonstrates the shortsightedness of the old saw that computers are dumb and can only do exactly what you tell them to do. Although in one sense the Creativity Machine "merely" follows the instructions programmed into it by Stephen Thaler, this does not mean that the Creativity Machine can only produce solutions Dr. Thaler expected or could have designed manually.[46] It might be tempting to think of the relationship between Dr. Thaler and the Creativity Machine as that between master and servant, but a better analogy, as Dr. Thaler himself puts it, is that between "proud parent and child" because he, like any good parent, tries best to point his progeny in the right direction and then set it free. Although Dr. Thaler created the process that the Creativity Machine uses to dream and tells the Creativity Machine what subjects to dream about, the machine did the rest when it produced the crossed-bristle configuration for the CrossAction toothbrush design. True, Dr. Thaler also supplied the Creativity Machine with the examples it needed of existing toothbrushes and performance data indicating how effectively each one brushed teeth. But such information did not include any expert knowledge or guidance about toothbrush design, and the Creativity Machine took the next crucial step: gleaning from that set of pure, unrefined data an understanding of what makes a great toothbrush great.

Computers do not, therefore, always operate according to the adage "garbage in, garbage out." Artificial invention technology, given only an abstract problem definition and simple rules for generating and evaluating possible solutions to the problem, can follow those simple rules to discover complex rules and patterns that its human programmer never imagined.[47] Just as we find in biological evolution that you can put primordial ooze in and get sentient life out, so too with search-based artificial invention technology can you put common knowledge in and get inventions out.

Search-based strategies are not guaranteed to produce *any* useful results, just as attempting to find a needle in a haystack may leave you with only a face full of straw. Furthermore, search-based strategies often generate potential solutions randomly; running a search multiple times for a solution to a single problem may produce one answer one time, a different answer another time, and no answer at all a third time. Usually, the only way to know whether a search-based strategy will produce a useful solution is to perform the search and observe the results.

In contrast, an automated design-based strategy is guaranteed to produce an answer to the problem you're trying to solve, at least in theory. If a programmer writes a program according to the rules of a programming language, we know ahead of time that a compiler will be able to fill in all of the details of that program "by design" automatically. Such success is guaranteed because the fact that a software compiler exists means someone else has already furnished the compiler with the mechanisms necessary to fill in the details of every instruction that can be written in the programming language. As a result, design-based automation can be useful when the extra effort required to specify the structure of the solution is justified by the benefit of achieving a detailed solution with certainty. In contrast, you might consider rolling the dice with search-based automation if the potential benefit of uncovering a solution that you could not have designed manually outweighs the risk that a search will not find any solution.

Although search-based strategies do not require you to know the *form* of the solution, they do require possible solutions to be evaluated. If you have no choice but to perform such evaluations manually, as Edison did with light bulb filaments, then executing a search-based strategy may be prohibitively expensive or even impossible. Performing tests automatically typically requires a *simulator* that can determine how a possible design—whether it be an antenna, toothbrush, or computer program—would perform under real conditions. Today there are highly accurate simulators for certain kinds of devices such as electrical circuits, but not for simulating less well-understood phenomena such as fluid dynamics. Because simulation is preferable to real-world testing, all other things being equal, artificial inventors are always seeking faster and more accurate simulators. Design-based strategies, in contrast, don't require simulators or testing of potential solutions.

In short, search- and design-based solutions have their own advantages and disadvantages, which require careful comparison case by case. Search-based solutions are adept at producing results that are good enough, but not necessarily perfect or applicable (1) across a range of circumstances,[48] (2) when you want to minimize the amount of human skill and effort that goes into the process, or (3) when you know what problem you want to solve but not what the form of the solution is.[49] Design-based solutions are good at producing results (1) whose inner workings can be understood, (2) for which we need a high degree of confidence that they will work across a range of circumstances, (3) when you have the time and other resources to dedicate human ingenuity to solving the problem, or (4) when you think you can at least divine the form of the solution beforehand.

But Is It Invention?

It is worthwhile explaining in more detail why I claim that the processes I have described qualify as automated inventing, particularly in light of two common objections to this claim. First, some argue that using design-based automation technology to automatically fill in predetermined details of a high-level outline written by a human does not qualify as automated inventing because deterministically filling in such details does not constitute "inventing." My claim, however, is not that a compiler, or other design-based automation mechanism, engages in inventing. Rather, my claim is that the *process as a whole*, performed by human and machine in combination, constitutes "automated inventing." Such a process amounts to inventing because it can produce inventions (as we saw with the antilock braking example). It is automated because part of the process is automated.

Such partial automation is significant for at least two reasons. First, even partial automation can create inventions the human inventor could not have invented manually, as the use of HDLs to design modern microprocessors demonstrates. Second, consider a case in which the wish (human-written abstract description) itself is inventive, as in the case of the antilock braking system we encountered earlier. When the inventor of the antilock braking algorithm applied a description of it to a design-based automation mechanism, the output was a new and useful machine capable of performing the algorithm. An electrical engineer who designed circuitry capable of performing the same algorithm to produce a new and useful machine would have engaged in an uncontroversial act of invention. Because such a machine is an invention in the sense that it is new and useful regardless of how it was designed, even deterministic design-based automation automates inventing in the sense that it automates the process of producing the physical design for the machine.

Second, some argue that search-based strategies, which involve randomness and trial and error, do not constitute inventing because such brute force methods lack the creativity essential to invention. A computer program that merely tries out a large number of possible designs and picks the best one according to criteria spelled out by a human, so the argument goes, does not invent in the same sense as a human inventor. My first response to this is that, once again, I do not claim that the software by itself invents. Rather, the human-computer system invents.

Second, search is a well-accepted technique of invention. Thomas Edison, perhaps the most prolific inventor of modern times, found the best light bulb filament of his day by hiring people to scour the globe to *search* for and test more than 6,000 candidate materials before *selecting*, not *designing*, the winner (carbonized bamboo).[50] Perhaps it was this kind of intensive search that led Edison to proclaim that "genius is 1 percent inspiration, 99 percent perspiration." If search cannot be sufficient for invention, then we had better strike the Patent Office records of much of its collection[51] and amend both the U.S. Constitution and the U.S. Patent Act, each of which considers "discoveries" to be inventions.[52]

It has been stated that the goal of artificial intelligence research is "to construct computer programs which exhibit behavior that we call 'intelligent behavior' when we observe it in human beings."[53] Applying the same standard to inventing implies that search-based automation techniques qualify as artificial inventing because we would consider a human to have engaged in inventive behavior by solving a problem through searching for possible solutions and finding the best one in the form of a design for a new and useful machine. Artificial invention technology has also proven its ability to pass this "inventive behavior" test by generating designs for which patents have been granted.

An artificial genie passes this test, in other words, if it can trick us into believing it is human. Peter Bentley of University College London (UCL) produced just such a trickster genie in the late 1990s, when he and a team of computer scientists and musicians from UCL and Brighton got together and made a secret evolutionary music composition machine that wrote music based on an automated analysis of Top 10 tracks of the day. The machine was secret because the major music studio—which shall remain nameless here—that commissioned the project didn't want the world to know that songs being released under its label and played in dance clubs were written by a computer rather than humans. The studio was so insistent on secrecy that it fronted human musicians to act as copyright holders and receive royalty payments. Bentley's team went on to found J13 Music, a record label, and J13 Technology, the company that owned the music-writing technology.[54]

The skepticism toward viewing search as a kind of inventing may have had a reasonable basis when searching was a prohibitively time-consuming and costly way to solve problems, unless you had Thomas Edison's wealth to devote to finding and testing thousands of possible solutions.[55] Now, however, a combination of improvements in search-based artificial invention technology and exponentially increasing computer processing speeds is undermining the basis for this skepticism and making search an ever more viable route for inventors lacking Edison's resources to invent. Recognizing this on an intellectual level is the easy part. The hard part is accepting, at an emotional level, that human ingenuity no longer stands at the pinnacle of inventive prowess.[56] Stephen Thaler likens this to previous generations' acceptance (at best grudging) of Copernicus' discovery that we were not at the center of the physical universe and Darwin's revelation that we were not the perfect result of a guided design process. If you know anything about the storms those two previous scientific revolutions caused, you have an inkling of what we may be in for now.

Many of the leaders in artificial invention technology whom I interviewed for this book related to me the difficulty they have in convincing both their peers and their potential customers that their technology can do what they claim it does. Although some of this is legitimate and healthy skepticism, it is often voiced explicitly as disbelief that a computer could ever outinvent a human. It is still common for scientists to proclaim that the human brain is the pinnacle of creation, the most complex and powerful product ever produced by nature.[57] Yet such claims are more ideological than scientific; they do not result from any rigorous comparison between the human brain and other complex organs or organisms.

What we need now, as we did in the times of Copernicus and Darwin, is humility about our place in the universe, and our relationship to the machines we create.[58] You will be no less human, creative, or valuable if you allow yourself to recognize the possibility that a computer can generate and evaluate many toothbrush shapes more quickly than you can. What you need to do to retain your sense of self-worth is shift your perspective; stop thinking of yourself as one of Henry Ford's engineers and start thinking of yourself as Henry Ford, who was proud to proclaim that his skill was not at knowing the answer to every question, but rather at knowing which questions to ask and having the best possible team of people at his side to tackle those questions for him. Start thinking about artificial invention technology as that team, as a supplement to, rather than a replacement for, your intellect and creativity, and you will realize that such technology can be a lever to raise yourself to even greater heights, rather than a hammer to beat you down. Although doing so may require you to give up some control, you will obtain greater power in exchange.[59] As we will now see, people who want to succeed in the Artificial Invention Age will need to know both themselves and the technology available to them, so that they can learn how to use a mixture of their own skills and the artificial skills of computers, to the best advantage of each.

6 Humans and Their Tools

Partners in Invention

IMPLICIT IN MUCH OF THE DISCUSSION so far is that the impact of automation technology on inventing is best understood if we (1) view human inventors and their tools as integrated inventive systems, working in partnership, rather than in isolation; and (2) recognize that computers can effectively *augment* the inventive skills of the human inventors who use them. If you've ever referred to Google or another Internet search engine as your "external memory" after using it to help you remember who starred in your favorite movie, then you already acknowledge that machines can act both *in concert with* and *as enhancements to* your mind.[1]

People typically use terms such as "human enhancement" and "transhumanism" to refer to modifications to the human body itself to overcome biological limitations, such as modifying our red blood cells to enable them to go for hours without oxygen,[2] taking a "pain vaccine" that will block intense pain in less than 10 seconds,[3] or even receiving a cochlear implant to improve hearing.[4] Although such alterations certainly are impressive, overemphasis on such direct bodily modification overlooks the more subtle and far-reaching reality: that what makes a search engine a "mindware upgrade"[5] is not that it is located within your biological "skinbag" but rather that it operates in close integration with your mental processes.[6] As Andy Clark, a University of Edinburgh philosophy professor, argues, we are cyborgs already even if our bodies lack the artificially implanted machinery of a Bionic Woman or Robocop.[7]

Not every human-machine interaction, however, produces a cyborg. For guidance on what makes certain human-machine systems special, both Clark and J.C.R. Licklider have looked to instances of biological symbiosis and concluded that the signposts of symbiosis are fluidity of interaction between two entities and tightness in their integration.[8] If interactions between a human and a machine satisfy such criteria, we have an instance of *human-machine symbiosis.*

A telling example of human-machine symbiosis is the one between airplane pilots and the planes they pilot. The need for military pilots to react not only to rapidly changing circumstances but to a vast array of flight deck instruments has spawned an approach to airplane design and pilot training that views airplane and pilot "as a single extended system made up of pilots, automated 'fly-by-wire' computer control systems, and various high-level loops in which pilots monitor the computer while the computer monitors the pilots."[9]

Similarly, whenever a human inventor and machine satisfy the criteria for symbiosis, the pair may profitably be thought of as a *single inventive system.* Although computer technology brings inventor-machine symbiosis to a new level, inventors have always used tools to assist them in inventing, with the sole exception of the first inventor, who used his bare hands to snap a branch from a tree and use it as a spear. Even that earliest inventor turned next to a rock to sharpen the tip of the spear, and thus began the use of tools as devices for helping to invent other tools.

Since then, every tool for cutting, molding, extruding, or otherwise bending matter to our will, from the axe to the hammer to the spinning wheel, has been used by inventors to build and test prototypes of their inventions more easily and effectively. So too have inventors used artificial calculation devices—from the abacus to the slide rule to the electronic calculator—to predict the weight, dimensions, and other physical properties of their inventions and therefore "test" them before building. The same is true of every writing implement any inventor has ever used to sketch an invention design, only to cross it out and start over after spotting a flaw. Every device that enhances our ability to observe the physical world, from the simplest lens to the electron microscope, greases the wheels of invention by enabling inventors to better observe and understand how their inventions behave and interact with the rest of the physical world. Inventors use tools as extensions of their hands, eyes, ears, and even brains to manipulate, observe, and analyze inventive works-in-progress, and thereby assist in understanding and improving those works more effectively than would be possible with the unaided human mind and body.

Invention Augmentation

These are just some of the ways in which human inventors and machines, working together, can *augment* each other's ability, resulting in an inventive system that is more powerful than either could be by itself. Computing pioneer Douglas Engelbart, who created the first computer mouse and window-based computer interface back in 1968, identified four primary kinds of tools—artifacts (e.g., physical machines), language, methodology, and training—for "augmenting man's intellect."[10] Part of Engelbart's brilliance is to recognize the augmenting power not only of physical artifacts but also of these other kinds of tools, and the relationships among them.[11] I will continue to use the term *tools* throughout the remainder of this chapter to refer broadly to both physical devices such as computers and the other "augmentation means" identified by Engelbart.

In general, a tool augments our intellect if, when you provide the tool "with a little intelligence, [it] emit[s] a lot."[12] Tools can augment our intellect in many ways, enabling us to:

- Understand or solve problems more quickly
- Understand or solve problems that previously we could not solve
- Produce better solutions to problems[13]

Furthermore, Engelbart recognized that all such tools can augment not just the intellect of the *human user* of those tools but more generally also improve the "system . . . visualized as comprising a trained human being together with his artifacts, language, and methodology."[14] Although word processors did not yet exist at the time Engelbart wrote his 1963 paper, he posited them as an example of a tool that could be introduced at a low level of the writing process to augment the process at higher levels. As author Howard Rheingold put it, Engelbart's "speculations sound like advertising copy for word processing systems of the 1980s," pointing out many of the benefits of word processing that we take for granted today.[15] In particular, Engelbart was careful to emphasize that the importance of the word processor is not its set of technical features in isolation but rather the "far-reaching effects" it has on one's creative abilities.[16]

These effects are particularly pronounced when the human-machine partnership is highly *interactive*. Try writing an essay longer than a few sentences in your head and you will see what I mean. Start writing the same essay on a scratchpad and you will find that the physical paper, with your work in progress stored on it in ink, serves as a temporary external memory to augment the significantly limited capacity of your onboard (biological) memory. As you "think" on the page, you also read back your own writing to prompt you to form additional thoughts and write them down as well. Physical writing is a two-way feedback loop between human and ink on paper, which enables you to formulate more complex ideas more clearly than you could if you had to rely on your unaided mind alone.[17]

Inventors throughout history have benefited from a similar kind of interactive feedback loop whenever they have built a prototype of a device, tested it, and then incorporated the results of their tests into an improved design. The current incarnations of artificial invention technology take the process to the next level first by enabling inventors to produce intermediate designs that surprise them, and from which they can gain new insights they can feed back into the inventive process, and second by enabling them to do so more quickly and easily than ever before. For example, the engineers who created the NASA antenna we encountered earlier used evolutionary algorithms to generate a first generation of antennas. They studied those preliminary designs and noticed that the strength of the signals the antennas produced at a given elevation varied over time, instead of being steady. In response to this observation, the engineers modified the fitness function of the evolutionary algorithm to favor antennas having smooth signal strengths. Then they ran the evolutionary algorithm again, generating better antennas as a result. The best antenna they eventually found was the result of such a human-computer feedback loop.[18]

This iterative process of forming hypotheses, building and testing prototypes, and learning from them belies the lockstep simplicity of the waterfall model I introduced earlier. Modern forms of artificial invention technology tend to facilitate and even encourage such a process, as James Foster of the University of Idaho told me:

> This happens all the time; that process of going back and refining the specification is a discovery process. [It] . . . helps you refine the problem, understand the problem better, and it may cause you to question assumptions which really

should be questioned, and wouldn't have been otherwise. . . . I'm a Jackson Pollock kind of guy. I like creating the art by splashing things on the wall and seeing what happens, and there is a taste of that in everything people do in [the evolutionary computation] community.[19]

Although this book is about the *automation* of invention, not about techniques for *augmenting* human skills generally, there is a close relationship between automation and augmentation. Automating one step in solving a problem can often augment a human's ability to solve the problem as a whole. We've already seen, for example, how automating the process of translating computer programs into working software augmented the programmer's ability to create more complex and powerful programs than before.

Another reason for focusing on augmentation of human skills by invention automation technology is that doing so puts the lie to the common assumption that automating human skills can only make human inventors obsolete. To the contrary, machines that automate human inventive skills can be used to augment those skills, thereby making the human skills more valuable than ever before.[20] We need to view humans and computers as potential partners in invention, instead of competitors or master and slave.[21] Even if we insist on analogizing computers to human body parts, Clark's "mindware upgrade" is far preferable to the traditional "electronic brain."

As late as the 1980s, when personal computers began to inhabit the desktop of every office worker and many home computer users, Doug Engelbart rightfully continued to bemoan the fact that even those at the forefront of the computer industry still failed to recognize how the "power of using a computer as a mind amplifier [lies] not in how the amplifier works but in what amplified minds are able to accomplish."[22] We must keep our eye from shifting lazily in wonder to the marvels of computer technology in isolation, lest we overlook the impact of such technology on human skills and the inventive process as a whole, in which it promises to impart "intellectual leverage of previously unimagined dimensions."[23]

Inventors Inventing Invention-Augmenting Tools: A Virtuous Cycle

Although inventors *use* tools to augment their abilities just as we noninventors use Google to enhance our memory, inventors have a leg up on the rest of us: they also *invent* tools to boost their inventive powers even further. I have seen this dynamic play out countless times in my work as a patent lawyer. Whenever I begin working on a patent application for a new invention, I interview the inventors about what their invention is, of course, but also about how and why they invented it. Typically, the first question I ask is, "What problem were you trying to solve when you invented the invention we're trying to patent?"

I've lost track of how many times the inventors have responded with a story of this form:

> We spend our days designing new kinds of X, and we found that every time we designed a new X we had to manually test a certain part of it. This involved figuring out which tests to perform, and then performing those tests manually. This was very tedious and time-consuming. Then we realized that we could write software that could design the tests for us and perform the tests automatically, both more quickly and accurately than we could. So we wrote the software, and now it saves us a lot of time when we are designing a new X, increases the accuracy of the tests we perform, and increases the quality of the resulting new X.

X can be anything that is capable of being simulated or tested by software. What all these stories have in common is that they involve inventors who have invented something *that helps the inventors themselves invent more quickly and with less effort.*[24] Computer programmers are notorious for their zeal in creating and adopting software they can use to reduce their own workload. A software package named Autoflow, arguably the first commercial software "product" in history, owed its resounding success to the ability to automate the programmers' task of writing flowcharts to document their software at the end of a project, "famously the most hated part of the programmer's task."[25]

Once inventors have invented a new invention-augmentation tool, they then use it to invent even more powerful invention augmentation tools, and so on, in a virtuous cycle. Engelbart, who coined the term *bootstrapping* to refer to this process,[26] founded and continues to run the Bootstrap Institute to promote use of the technique to accelerate innovation. As a real-world example of bootstrapping at work, Jeff Walker of TenFold related to me how the company used its own software-creating software to build the next generation of "TenFold Tools," the software that interacts with the company's customers to enable them to build software.[27]

Although inventors have always used bootstrapping, the latest generation of artificial invention technology enables inventors to take bootstrapping to the next level because such technology can *internally and automatically* generate, simulate, test, and refine potential designs, an endeavor that previously could only be performed by a human or a human-machine system. Figure 9 shows how inventors traditionally have used tools to assist them in *building* prototypes once the inventors themselves have designed those prototypes, while Figure 10 shows how today's artificial invention technology goes further by automatically and iteratively generating, evaluating, and modifying *designs*

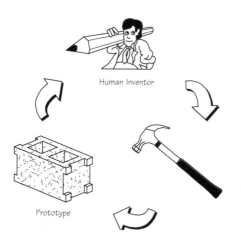

for the prototypes themselves. Even if a computer-generated design is not perfect and the human inventor's input is required to improve it, the use of a computer to perform even an initial separation of the wheat from the chaff reduces the extent of the human inventor's involvement in low-level design, with all of the attendant benefits we've already seen.

FIGURE 9 Manual design and evaluation. Courtesy Spinney Associates

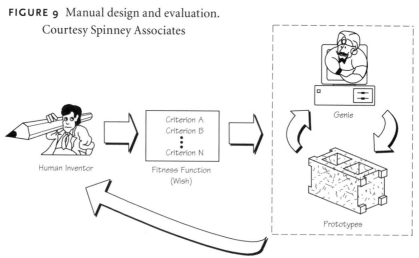

FIGURE 10 Automated design and evaluation. Courtesy Spinney Associates

Secondary Effects

Although we've seen how inventors and their tools work together at a particular point in time, neither humans nor tools are fixed; they influence each other and thereby *coevolve* over time.[28] In particular, humans develop new tools, which influence the mental and physical processes that humans apply to those tools, which in turn influence the development of yet further tools. Consider the coevolution of humans and automobiles. Human experience with horse and buggy influenced the designs of early automobiles. Human experience with early automobiles then prompted the addition of new features (such as windshield wipers and air conditioners) not found on any horse-drawn carriage. Furthermore, experienced human drivers developed faster reflexes, in response to which automobile engineers designed faster cars.

These are only the primary effects of humans and tools on each other. Those of us who were born into a nation of automobiles, including a national infrastructure developed around an extensive road system with gasoline filling stations and the services needed to keep cars running smoothly, now plan our daily routines—and our lives in general—around the assumption that cars and roads are readily available. We are willing to live 50 miles from a city, trusting that we can commute daily, quickly and cheaply. We call a friend in the next town to schedule dinner an hour later, whereas in the absence of cars and roads such a meeting would have had to be scheduled and carefully planned days or weeks in advance. Tools influence not only human skills but also human attitudes and habits.

Such dynamic feedback between humans and their tools over time is particularly pronounced in the context of inventors and their tools, precisely because inventors have the will and the skill to consciously invent tools designed specifically to augment their own inventive abilities. This gives inventors the power to push forward their coevolution with their tools more rapidly and with more focus than the rest of us, who must rely on whatever tools inventors have made available to the public. One consequence of this is that skills previously required by inventors quickly become unnecessary for inventors even a few years hence. Programmers who grew up in an age of assembly language never needed to learn how to program plugboards by hand; those who grew up with automated compilers did not need to learn how to program in assembly language. Inventors who use today's more advanced forms of artificial invention technology may never need to learn how to engage in the kind of structural design and direct physical experimentation that has been the stock-in-trade of inventors literally since the beginning of human history. In this way, inventors develop only those skills necessary to leverage the current state of the art in invention augmentation technology. From where each generation of inventors stands, the tools at its disposal are the state of nature, to be used as stones to shape spearheads and continue the coevolution of humans and their tools.

II REGULATING GENIES

7 Patenting Artificial Inventions

NOW THAT WE'RE FAMILIAR with how artificial invention technology works, let's examine how patent law should operate in the Artificial Invention Age. First, however, we must explore how patent law is intended to promote innovation, and why simply letting the legal status quo play itself out will *not* promote innovation in the Artificial Invention Age.

Patent Law Basics

A patent on an invention grants the inventor exclusive rights to the invention in the sense that the patent empowers the inventor to exclude others from making use of the invention.[1] Patents are often considered a kind of "intellectual property" because they bestow on their owners a kind of property right in intellectual goods—inventions—distinct from the particular physical form those inventions may take, such as specific physical light bulbs, cotton gins, and software. These exclusive rights have an important limitation: they don't last forever. The U.S. Constitution states explicitly that copyrights and patent rights are to be granted only "for limited times."[2] For example, a U.S. patent granted today is enforceable only for 20 years from the date on which the patent application was filed.[3] If you think of a patent as a kind of property, then limiting its term (duration) may seem strange; government agents don't come to your house and take away your lawn mower after 20 years. We limit patent rights in this way to provide an incentive to inventors to create new inventions and then make those inventions available to the public. Patent rights do this by giving inventors a reasonable assurance (not a guarantee) of a reasonable return on their "inventive investment"—the time, effort, and money they put into the act of inventing.[4]

Imagine that you are an automotive engineer about to embark on designing a more aerodynamic automobile frame. You estimate that it will take a year of your time to do so, which you value at $50,000 in labor. You're trying to figure out whether it will be worthwhile for you to devote your time to such a project. I'm a representative of the public, and I come to you and say, "How much would we need to pay you at the end of the year to make it worthwhile for you to spend that year working on the frame design?" "$75,000," you respond, taking into account that you won't get the money until the *end* of the year. Then I say, "We can't *guarantee* that we will be able to pay you. The best we can do is give you a *reasonable assurance* that we will pay you. Now how much do you need?" Taking into account the risk that you will spend a year in the workshop only to receive nothing in return, you respond "$250,000."

The exact dollar amount of your answer isn't important. What is important is that there is some *finite* amount of money that will satisfy you. If we, as the public, offer you too little then you'll say "no deal," and if we offer you too much you'll walk away with a windfall, leaving us with less money to pay other inventors. The same logic applies to the term (duration) of the patent; it should be made just long enough to encourage inventors to invent and disclose those inventions to the public. Patents are often viewed as just this kind of contract between inventors and the public, in which the inventor agrees to disclose the invention to the public in exchange for exclusive rights to the invention.[5]

If we agree that patents can be useful for bringing new inventions to the public, we still need a way to determine whether any particular invention deserves a patent. Once inventors get wind of the fact that they can lock out their competitors by obtaining patents, every Tom, Dick, and Harry with a mousetrap in hand will emerge from the woodwork in an attempt to patent it. We, as the public, therefore need to enact strict rules to govern what can be patented. For example, we need to prevent patents from being obtained on inventions that are *already* available to the public. There is no reason to grant a patent on a mousetrap that has already been available for sale at the local five and dime for years; the reason for granting patents is to encourage inventors to disclose to the public what those inventors would otherwise keep secret. This is the purpose of patent law's *novelty* requirement, which ensures that only inventions differing in some way from existing technology can be patented.[6]

We also need rules to ensure that patents are granted only on inventions that would not have been invented and made available to the public in the absence of a patent system.[7] Patents are needed only in situations in which market forces, standing alone, would *fail* to produce an invention and make it available to the public. But markets don't always fail in this way. The sheer number of products you can buy that are not patented demonstrates that markets alone are often sufficient to produce innovation.[8] For example, if an inventor can make a minor improvement to an existing mousetrap both quickly and cheaply, then the inventor may be able to make a profit on that mousetrap, even without a patent on it, by bringing the improved mousetrap to market quickly and selling it in large numbers before competitors can enter the market. Granting a patent on the mousetrap in this situation may be unnecessary or even contrary to the public interest.

Patent law uses what is known as the *nonobviousness* requirement in an attempt to weed out inventions that would likely be created and made available to the public even without patent protection.[9] Although it is dangerous to attempt to summarize what the nonobviousness requirement means, one way you can think about it is as a way to weed out inventions that resulted from making only trivial modifications to previous inventions, such as merely changing the length of an existing car frame by an inch or substituting one kind of metal with another in its construction.[10] Intuitively you might view such modifications as "trivial" or not sufficiently "inventive," and understand the nonobviousness requirement as a way of preventing patents from being granted on inventions that, although technically new, differ from previous inventions only in a trivial way not worthy of patent protection.

You can think of the function of the nonobviousness requirement in another way. If any ordinary automotive engineer, working away at the daily grind, would have found it obvious to increase the length of a car frame by an inch, then we don't need to grant a patent to you if you happen to be the first person to make such a modification.[11] Even if the unavailability of patent protection for your car frame causes *you* to decide to keep it a secret, another automotive engineer will likely come up with the same idea tomorrow and incorporate it into an automobile made available to the public. Therefore we, the public, don't need to grant you a patent on the inch-longer car frame in order to ensure that it will become available to us.

Finally, we don't want to grant patents on laws of nature (such as the law of gravity), natural phenomena (such as electricity and wind), or abstract ideas. Patent law won't let you obtain a patent on any of these things, even if you're the first to discover them.[12] You can, however, obtain a patent on what is called a *practical application* of a law of nature, natural phenomenon, or abstract idea.[13] For example, gravity is a law of nature and therefore unpatentable, but a weightlifting machine is a practical application of the law of gravity and therefore patentable. $E = mc^2$ is an abstract idea; a particle accelerator is a practical application of that idea and therefore patentable.

When you attempt to obtain a patent on your new and improved automobile frame, the burden is on you to prove that the frame satisfies these three requirements of novelty, nonobviousness, and utility. You do so in countries worldwide by writing a description of your automobile frame in a document called a *patent application* and submitting the application to a government patent office for review. Your patent application must describe how to make and use your automobile frame in a way that would enable a typical auto worker to build a working version of the frame. One function of the patent application, in other words, is to act as a blueprint that others can follow to make and use your invention. This is referred to as the *public disclosure* function of patents, and the various requirements mandating that you describe in your patent application how to make and use your invention are referred to as the *disclosure* requirements.[14]

The patent office examines your patent application to determine whether the invention it describes is new, useful, and nonobvious and therefore qualifies for patent protection. However, in an attempt to save its employees unnecessary effort the patent office requires that you point out specifically in your patent application what makes your car frame special. You do this by writing additional text referred to as *claims*, because they indicate what you claim your invention to be. For example, you might write a claim pointing out the unique shape of your car frame, which makes it so aerodynamic. Whereas the rest of the patent application, usually referred to as the *specification*, may go into great detail about how to *build* your car frame out of metal, the claims cut to the chase, pointing out the essential features of your frame that distinguish it from all other frames that came before it.[15]

Assuming that your patent application passes muster, it will be granted and become a patent. Just as the text of a bill becomes a law, the text and drawings that you submitted in your patent application become part of the final patent itself. The patent office publishes the patent for all the world to read. This again promotes the goal of using the patent system to make inventions available to the public, because anyone who understands mechanics can read your patent to learn how to build your car frame. Of course, people must wait until the patent expires about 17 years later if they wish to avoid being slapped with a patent infringement lawsuit by you. In the meantime, however, any member of the public, including your competitors, can read your patent to study your car frame. Much to your chagrin, this might give your competitors some bright ideas about frame design, which might lead them to invent an even better frame, which they might patent themselves. Although this may not be good for you, it is good for the public because it leads to creation of another patented invention.[16] Once your patent expires, anyone can make and use your frame even without your permission, at which point the patent system once again has served its purpose of making more inventions available to the public.[17]

While your patent is in force, you have two choices for making money from it. First, you can manufacture your car frame and sell it, commanding a premium price because of your ability to exclude your competitors from the marketplace. Alternatively, you can sell the patent outright or charge other companies a fee (royalty) for the right to manufacture and sell the car frame themselves. Without the patent, you would have little or no leverage to extract such payments from them; they could just manufacture the car frame themselves without paying you a dime. With the patent, however, the manufacturer knows that they will face a patent infringement lawsuit if they sell the car frame without your permission.[18] In such a case, everyone is happy: you, because you've made money from your patent without having to manufacture anything; the manufacturer, because they've made money from your patent without having to invent anything; and the public because it can now drive down the highway with less air drag than ever before.

Why Existing Patent Law Will Produce the Wrong Answer

> There are no hard problems, only problems that are hard to a certain level of intelligence. Move the smallest bit upwards [in level of intelligence], and some problems will suddenly move from "impossible" to "obvious." Move a substantial degree upwards, and all of them will become obvious.
> —*Eliezer S. Yudkowsky*[19]

Now that we have a basic understanding of how patent law works, we can return to figuring out which rules patent law should follow to determine whether any particular artificial invention—car frame, toothbrush, or software—is patentable. I will begin with an example, demonstrating that patent law's *nonobviousness* requirement will break down if it is applied to artificial inventions. As a result, the nonobviousness requirement will need to be re-interpreted in the Artificial Invention Age if nonobviousness is to continue serving its intended purpose.

Patent law considers a particular invention to be obvious if a "person having ordinary skill" in the field of the invention would have found the invention obvious at the time the invention was made. (Patent lawyers use the term "art" rather than "field," and as a result we talk about a "person having ordinary skill in the art," or simply PHOSITA.) Although this definition is notoriously difficult to apply, in general it means that a particular car frame is obvious, and therefore not patentable, if an automotive engineer of ordinary skill would have found it obvious to design such a car frame at the time the car frame was created.

We've seen, however, that artificial invention technology is changing the effective level of skill of ordinary inventors in a variety of fields. Supply every automotive engineer with state-of-the-art artificial invention technology and train them in how to use that technology, and you have effectively boosted the level of ordinary inventive skill in the field (art) of automotive engineering. If instead all of those automotive engineers had gone back to school and obtained a Ph.D. in automotive engineering, the law would recognize that the level of skill of the ordinary automotive engineer had just increased a notch for purposes of applying the nonobviousness requirement. This would raise the bar for nonobviousness in the field of automotive engineering, thereby making it more difficult to obtain patents in the field.

Yet the law today does not explicitly take into account the introduction of improved invention augmentation technology in the same way as it takes into account improvements in education for purposes of identifying the level of ordinary skill in a particular field (art).[20] The result is that the nonobviousness bar may remain lower than it should be in fields in which invention augmentation technology becomes commonplace, and inventors in those fields who leverage such technology will be able to overcome an artificially low nonobviousness threshold, thereby obtaining patents that should be rejected. Such patents could have the perverse effect of impeding rather than promoting innovation.[21]

To see how this scenario would play out in practice, imagine you're a college student studying automotive engineering. During your first semester you learn a mathematical formula that describes the most efficient way for air to flow around the frame of an automobile.[22] Although the equation has been well known in the field of automotive engineering for decades, professional automotive engineers continue to come up with new and clever designs for automobile frames consistent with the formula. Coming up with a new frame design that satisfies the formula, however, isn't easy; it takes a combination of engineering skill, intuition, and luck. It's exactly the kind of structural design skill you went to engineering school to acquire.

Your final assignment in the class is to design a new automobile frame whose aerodynamic properties are consistent with this well-known formula. You get to work. After a month of effort, hundreds of sketches of failed designs, a dozen prototypes, and a combination of insight and caffeine, you come up with a design that no one has conceived before. You plug in the design to a computer simulator and, lo and behold, virtual air flows around it in the manner described by that old airflow formula.

You show your design to an automotive engineer. To your great surprise and initial disappointment, she tells you that your car frame just won't work—the frame violates the design rules she has learned and the intuitions she has developed as a professional engineer. But you show her the airflow simulations, she crunches the numbers, and eventually she is convinced that your frame *does* work, even though she still doesn't quite understand *why*. Somehow, as a first-year engineering student, you have managed to break free of the strictures of your future field. Pleasantly surprised and encouraged by your meeting with your engineer friend, you submit a patent application for your design before you turn in the assignment to your professor.

No one would question that you are entitled to a patent in the story as I've told it. Although the airflow formula is well known and not patentable itself, and although the formula describes what your automobile frame *does*, your frame is more than the formula. It is the kind of practical application of the formula that can be patented even if the formula cannot. If this weren't true, *nothing* would be patentable; every useful machine and process operates according to the unpatentable laws of nature.

Now transport yourself to an alternate universe in which artificial invention technology has advanced by leaps and bounds and is widely available at low cost. You again find yourself studying automotive engineering and again face the same end-term assignment: to build a better car frame. This time, however, instead of hauling yourself to the workbench you sit down at your desktop computer and feed that old airflow formula into the artificial invention software that you downloaded for free over the Internet. You click "Invent!" and watch a late-night movie. Before you can finish the first bag of microwave popcorn your computer spits out a design for a car frame—exactly the same frame that took so many hours of labor and so much pulled hair to invent in your alternate life.

Your intuition probably tells you that your artificially invented car frame is no longer patentable, even though nothing about the car frame itself has changed. Although your intuition is correct, we need to be careful to identity the right justification for the intuition. The reason your car frame should not be patentable is *not* that you invented it using artificial invention technology. Rather, your car frame is not patentable because *anyone having ordinary skill could have invented the same frame using artificial invention technology.* The relevant difference between the two scenarios we have explored is not, in other words, that *you* invented the car frame manually in the former instance and automatically in the latter, but that in the latter scenario you invented the car frame in a world populated by publicly available artificial invention genies.[23]

In such a world, inventors equipped with artificial invention technology have a different, possibly higher, effective level of skill than inventors in the same field who lack such invention augmentation technology. By the "effective" level of skill of an inventor, I mean the level of skill of the *inventive system* comprising the inventor and the invention augmentation technology that he or she typically uses. If an automotive engineer with a bachelor's degree who uses artificial invention technology to invent can compete head-to-head against a Ph.D. engineer who *doesn't* use such technology, then the inventive system consisting of the bachelor's degree engineer and the artificial invention technology has the same effective level of skill as the Ph.D. engineer working alone.

This point is worth emphasizing, because some of its consequences are unintuitive. Consider this twist: you are still an automotive engineering student in the Artificial Invention Age, but you remain stuck in a 20th-century mind-set and invent your car frame the old-fashioned way, *without* using artificial invention technology. Recall that in a genie-free world we considered the car frame design to be nonobvious and therefore patentable. That same car frame design is obvious now that you have invented it in the Artificial Invention Age, because even though *you* didn't invent it using artificial invention technology, the design would have been obvious to an automotive engineer having ordinary skill, because such engineers in the Artificial Invention Age know how to produce new automotive designs by inputting well-known airflow formulas to artificial invention technology.

Whether the car frame you invented is obvious does not, in other words, depend on whether *you* invented that frame manually or automatically. What matters is whether you invented the car frame *in a world populated by genies*, because such genies raise the background level of ordinary skill that informs what is obvious. Such a higher nonobviousness threshold is applicable to *all* inventions, whether invented automatically or not.

Interpreting nonobviousness in this way is consistent with the purpose of the nonobviousness requirement: to ensure that patents are not granted on inventions that would be made available to the public even in the absence of patent protection. If any automotive engineer having ordinary skill would find it obvious to invent a new car frame by handing the old airflow equation to an artificial genie, then there is no reason to grant *you* a patent on such a car frame, regardless of how you invented it. By putting the car frame design within the reach of anyone with ordinary skill, artificial invention technology makes granting a patent to you on that car frame not only unnecessary but positively harmful to the public interest.

The Problem with Existing Law

Patent law as it exists today is at best unclear about whether the nonobviousness requirement[24] must be interpreted to take artificial invention technology into account in examining patent applications.[25] In fact, there is good reason to believe that if you submitted a patent application for an artificially invented car frame today, your patent application would sail through the Patent Office with flying colors—at least if you pursued the patent as follows.

You write a patent application which describes the car frame design produced by artificial invention software after you provide it with the airflow equation. You write this patent application to look just like an old-fashioned patent application on a car frame. It includes drawings of the frame shown from various angles, explains how to manufacture the frame, and points out its beneficial aerodynamic properties. You don't say a word about the fact that you produced the car frame design by giving the airflow equation to an artificial invention genie.

As a result, anyone who reads the patent application—including the patent examiner at the patent office—would have no way of knowing that you did not invent the frame the old-fashioned way in a workshop. You are not necessarily required to blow the lid on your little secret; the law requires only that you explain how to *make and use* your invention—which you have done—*not* how you *invented* the invention.[26] The U.S. Patent Act even includes a provision stating that the patentability of an invention shall not be negated by the manner in which the invention was made (invented).[27]

Next, you submit the patent application to the Patent Office. The patent examiner searches through existing car frame patents and cannot find any that are similar to yours. As a result, the examiner decides that your invention is new and therefore satisfies the law's "novelty" requirement.

The patent examiner applies the "nonobviousness" requirement and determines, quite correctly, that your car frame would not have been obvious to an automotive engineer of ordinary skill. Remember that the frame has a shape that is surprising to automotive engineers and violates the principles you learned in engineering school. This makes it a classic case of a nonobvious design. But little does the patent examiner know you were able to produce this surprising result *merely by applying ordinary skill* because you used software that doesn't shy away from trying designs human engineers of ordinary skill have mental roadblocks against trying.

Your patent application also satisfies the "enablement" requirement because the written description of the frame that you included in the patent application includes sufficient detail to enable a person having ordinary skill in the art to make and use the car frame. After all, you've included drawings of the frame and explained the materials out of which it is constructed and how to manufacture it. Your patent application satisfies the "patentable subject matter" requirement because the car frame is a kind of machine. The patent application satisfies the "utility" requirement because your car frame's aerodynamic features make it useful.

The result is that you obtain a patent, retire to Bermuda, and grow rich on royalty checks. Not a bad life.

Something is wrong here. You've obtained a patent on an invention by complying with the letter of the law, even though you've violated the spirit of the law (intentionally or unintentionally) because your invention is obvious. The best proof of this is the fact that you only used ordinary skill to invent it. This is precisely the result that patent law's nonobviousness requirement is intended to prevent.[28]

This breakdown results from the fact that the patent examiner considered only whether the car frame would have been obvious to a *human engineer, using only his personal skill and 20th-century inventive technology,* when in fact the available artificial invention technology made it possible to use only *ordinary* skill to invent the car frame.[29] As we know, inventors do not invent using their minds alone; they use all of the tools at their disposal.

> It is likely that each individual develops a certain repertory of process capabilities from which he selects and adapts those that will compose the processes that he executes. This repertory is like a toolkit. Just as the mechanic must know what his tools can do and how to use them, so the intellectual worker must know the capabilities of his tools and have suitable methods, strategies, and rules of thumb for making use of them. All of the process capabilities in the individual's repertory rest ultimately on basic capabilities within him or his artifacts, and the entire repertory represents an integrated, hierarchical structure.[30]

The patent examiner reached the wrong conclusion because he did not take into account the inventive skill-boosting effect of publicly available artificial invention technology on the inventor of ordinary skill. Applying such a hobbled nonobviousness standard in the Artificial Invention Age could allow a small number of opportunists to obtain countless patents on obvious inventions. This is known as a "patent flood."[31] The problem with patent floods is that they can concentrate legal and economic power in the hands of a select few, not because they have engaged in a greater amount of legitimate inventive activity than others, but merely because they happen to be savvy enough and have the resources to get through Patent Office's front door first. Those early patent holders can then use the power that patents confer to extract rents from legitimate innovators.

Fortunately, nonobviousness can be fixed by breaking it free from its Industrial Age moorings. Consider the famous patent case of *In re Winslow,*[32] which involved a patent related to plastic bags. In the case, Judge Giles Rich determined whether Winslow's plastic bag was obvious by traveling back to the time when Winslow was sitting in his workshop, working on his invention with all existing plastic bag patents hanging on the walls around him. Judge Rich asked us to imagine that Winslow saw an existing bag holder patent, by an inventor named Gerbe, and realized that the bag holder could be improved to do a better job at holding the flaps of the bag closed. Winslow might have asked himself, "Now what can I do to hold the flaps of the bag closed more securely?" "Looking around the walls" filled with patents, Judge Rich continued, Winslow would have seen another existing patent by someone named Hellman, which described a mechanism that could be incorporated into Gerbe's bag holder to make it work better. As a result, Winslow might have imagined a new bag holder that incorporated features from both Gerbe's and Hellman's patents.

Because this combination of features was precisely the solution for which Winslow was attempting to obtain a patent, Judge Rich concluded that Winslow's invention was obvious. In other words, Winslow's invention was obvious and therefore unpatentable because it was merely the result of combining together the teachings of two existing patents in an obvious way.

This way of thinking about obviousness conceives of invention as a purely mental act; inventors invent by absorbing information about existing inventions from existing patents and other sources, and then simply *pondering* potential modifications to those inventions. They do not use external tools to aid their understanding, build and examine prototypes, or conduct experiments. There is some intuitive appeal to this conception of invention. Want to make your car go faster? Then search through patents on engines from other vehicles, such as jet boats and motorcycles, and try to find one that would work in your car, possibly by combining some of its features with features of engines from other patents.

Such a framework may have been accurate enough for patent law in Winslow's day, but it has grown so far out of step with how inventors work today that it is no longer sufficient as a foundation for the nonobviousness standard in the Artificial Invention Age. A view of nonobviousness in which an invention is considered obvious only if a person having ordinary skill in the art would have *thought* to make a particular *physical* modification to existing inventions fails to recognize that inventors no longer are limited to inventing devices by *conceiving* of the physical structures of those devices, or even by *building* those devices and testing them themselves. You didn't invent your car frame by forming a mental picture of it, drawing it, or building one from metal. You didn't take shapes from existing car frames and mentally combine them together. Instead, you invented your car frame merely by presenting an abstract description—a wish—of the *problem* you were trying to solve to artificial invention software, which then used the wish to produce the resulting car frame. If the software you used is publicly available, then anyone having ordinary skill could have used that skill to produce the same result as you. Patent law's nonobviousness requirement must be able to take this into account if it is to avoid concluding that your car frame is nonobvious, and therefore patentable, when in fact it is not.

The Future Is Here

Lest you think that everything I've said so far reads like a science fiction novel, rest assured that what I'm describing will come to pass. It has *already* come to pass in the context of wishes version 1.0: computer programs. Recall that old-fashioned software is the earliest incarnation of an artificial invention. You create a piece of software by writing a wish in the form of a computer program. You give that wish to a computer, which, acting as a genie, grants your wish by creating functioning software that can do whatever your wish described.

If software is already a kind of artificial invention, then we should expect the kinds of problems I've described to arise whenever we attempt to apply patent law's nonobviousness requirement to software. In fact, such problems have arisen, or at least so argue countless commentators since the 1980s. The problem is difficult to measure precisely because, if it is real, it afflicts both patent offices and the courts that are called on to review the validity of patents. Software patent critics, however, have long argued that software patents in particular are granted routinely on "obvious" or "trivial" developments.[33] Although such critics don't characterize their arguments in the terms I've used here, at bottom we are pointing to the same problems resulting from the same root causes.

One common criticism, for example, is that patent examiners routinely fail to find or understand the relevant *prior art*—such as previous patents on similar inventions—in attempting to determine whether a particular software patent application covers an obvious development. One reason for this, at least in the early days of software patents, was that patent examiners did not have access to good databases of software patents or academic software journals, and that patent examiners with an electrical engineering degree rather than a computer science degree were assigned to examine software patent applications.[34] It is not surprising that patent examiners in such circumstances would be at a disadvantage in finding evidence of obviousness and as a result might find patent applications to be nonobvious more often than was warranted. Another way to interpret this result is that a patent examiner with a degree in electrical engineering would tend to consider whether an *electrical* engineer, applying the ordinary skill of an *electrical* engineer, with the tools normally available to such an engineer, would have found it obvious to produce the software being claimed. Such patent examiners would fail to find certain software patent applications to be obvious for exactly the reasons I've presented herein.[35]

As another example, when the flood of software patents began in the 1990s, many argued that patents were being granted on the mere "computerization" of existing well-known techniques, such as Amazon.com's "one-click shopping" patent and Priceline.com's "reverse auction" patent. Critics of such patents argued that although it may have been true that no had ever used a *computer* to engage in one-click shopping or to hold a reverse auction, the "offline" versions of such methods were well known, and anyone with ordinary *programming* skill would have found it obvious to perform such methods using a computer. This again parallels my argument that merely applying a well-known airflow formula to a new artificial genie should be considered obvious, and that if we fail to prepare for this problem, patent offices will get the answer wrong.[36]

In summary, traditional software patents have already raised the problems the next generation of artificial invention patents will elicit for patent law, albeit to a lesser degree. To the extent that software patents have served as a test case for artificial invention patents, the results are not encouraging. We have reason to believe that the problems with nonobviousness will only become more pronounced as artificial invention technology improves and enables inventors to create inventions using increasingly abstract wishes.

Acknowledging Genies

The solution to this problem is straightforward—if not obvious (pun fully intended)—at least in theory: untie the patent examiner's hands. To determine whether your car frame is obvious, we should update Judge Rich's allegory by asking whether an automotive engineer sitting in a room with patents on a wall, physics texts on her lap, *and a computer on her desktop equipped with the kind of artificial invention software commonly used by inventors in the same field* would have *produced*, not just *thought of*, your car frame merely by applying ordinary skill to that software.[37]

Updating the nonobviousness requirement for the Artificial Invention Age in this way acknowledges that human inventors routinely use particular kinds of tools in the inventive process, and that humans and their tools work together as a human-machine inventive system. It also produces a result that comports with our initial intuition, namely that your car frame is obvious and therefore not patentable because an automotive engineer of ordinary skill would have produced the same frame design by inputting a known wish (the old airflow equation) to well-known artificial invention software. If artificial invention technology effectively makes engineers in a particular field *more skilled* at inventing, then the nonobviousness standard should reflect such an increase in the effective level of ordinary skill of engineers in the field. The result would be to raise the nonobviousness bar that inventors in the field have to hurdle if they want to obtain patent protection.

We've come full circle to produce a legal test for nonobviousness that is consistent both with our intuitions and with the purpose of the nonobviousness requirement, which is to prevent people from obtaining patents on inventions whose creation requires only ordinary skill.[38] Furthermore, the test is flexible, capable of adapting to changing technology over time, because it requires that we ask whether a particular invention is obvious by reference to how inventors in a particular field, at a particular time, actually engage in inventing.[39]

Nonobvious Consequences of Updating Nonobviousness

It is worth repeating that if artificial invention technology raises the bar for nonobviousness, it will do so not just for patent applications on "artificial inventions"—inventions produced using artificial invention technology—but for *all* patent applications on inventions in fields in which inventors of ordinary skill commonly use artificial invention technology. Therefore, although we began this chapter by asking whether *artificial inventions* should be patentable, we have found that the increasing power and ubiquity of artificial invention technology will affect the nonobviousness standard as it applies to *all* inventions, not just the artificial.

To bring this back to present-day reality, if today's automotive engineers of ordinary skill all follow 10 well-known design rules in trying to invent new car frames, then when *you* apply for a patent on a new car frame it had better not be one that would be produced using those well-known design rules, even if *you* did not use any of those rules and are completely ignorant of them. The state of the art in a particular field sets the threshold for obviousness, even against inventors who lack the knowledge or resources to take advantage of the state of the art. To do otherwise would allow inventors who are ignorant (perhaps willfully) of the state of the art to obtain patents on inventions that were not obvious to *them*, even though such inventions would be obvious to people having ordinary skill in the same field.[40] Granting patents to the least knowledgeable inventors can't possibly be the right way to interpret nonobviousness.

This poses a serious problem for a patent examiner tasked with examining a patent application for a car frame in the Artificial Invention Age. In reviewing such a patent application, the patent examiner will not need to know whether the car frame covered by the patent application was invented using artificial invention technology. Instead, the patent examiner will need to determine both what kind of inventive *skill* and what kind of inventive *technology* are commonly used in that field, and then apply the appropriate level of effective skill in the art to the obviousness inquiry as applied to the car frame in the patent application under examination.

What this means is that the patent examiner will need to ask whether an inventor of ordinary skill, *using artificial invention technology,* would have found the invention at issue obvious. One way to do this is to ask whether *applying well-known wishes to commonly used artificial invention technology would have produced the invention* under examination. Although the examiner's direct task is to determine whether a particular invention is patentable, to do this job properly requires asking whether any *wishes* that could be used to *produce* the artificial invention are obvious. In the case of our car frame, if the examiner can find an airflow formula (wish) that is well known to automotive engineers, then the resulting car frame is likely to be obvious.

Therefore, although in principle the law of nonobviousness needs only updating rather than fundamental change, the implications for the day-to-day administration of the patent system by patent examiners (and courts) are much more significant. To examine our car frame patent application, the patent examiner will have to search not only through existing patents on car frames but also through physics texts, mathematics journals, and any other source of possible wishes that could be applied to artificial invention technology to produce the car frame under examination. In Chapter 10, we will explore how this and other side effects of artificial invention technology will push the patent system to its limits and possibly beyond.

8   Follow the Value

WE HAVE JUST ASKED AND ANSWERED the question, "Should inventions produced using artificial invention technology be patentable?" Next we turn to the question, "Should wishes be patentable?" In other words, should an abstract description of a problem to be solved be patentable, provided that a computer is capable of producing a concrete solution to the problem on the basis of the description?

This may seem like a strange question, and one that only a patent lawyer would even consider asking. How could a mere *description* of a problem possibly be patentable? After all, the U.S. Court of Appeals for the Federal Circuit has held that defining a chemical solely in terms of its principal biological property constitutes only "a wish to know the identity of any material with that biological property," and that merely stating such a wish in a patent is insufficient to warrant patent protection for the chemical.[1] More generally, patent law is intended to apply to physical products, such as machines or chemicals or clothing, and even to industrial processes, but not to written or spoken words. Because the very idea of "wish patents" may strike you as bizarre, let me begin not by arguing for why they would be a good idea but by demonstrating once again that if we do nothing to reinterpret patent law, wishes nonetheless will become patentable simply as a logical outgrowth of how patent law works. Then I conclude that since wishes will become patentable in the absence of action to the contrary, we'd better start taking action now to make sure that patent law handles wishes in a way that promotes innovation.

The best evidence that wish patents could exist is that they already *do*; they're called software patents. Read a patent on software for clarifying x-ray images and what you will see is a description of what the software *does* and perhaps a description of how to *write* the software. Software patents, in other words, furnish programmers with instructions for how to program a computer to create the software covered by the patent. They tell programmers how to make a wish to a computer for that software. Such patents satisfy the legal requirement mandating that a patent enable the public to *make* the invention differently from most patents on mechanical devices, which show and describe the *physical structure* of the invention, whether it be a gear, toaster, or automobile engine. In contrast, a patent on x-ray clarification software does not typically show and describe the physical structure of the software itself, such as by presenting a complete binary listing of the software's machine code (which would indicate the sequence of on-off positions of switches in a computer's memory when programmed with the software). In a lawsuit over an early software patent, the defendant argued that the now-accepted style of writing software patent applications—describing software merely in terms of instructions for generating the software—was impermissible because it failed to point out the particular physical structure of the software covered by the patent. The court rejected this argument and held that describing how to *create* the software was sufficient.[2] Software patents have been written in this "here's what you need to do to *create* the software" way ever since.

Software patents came to be written as instructions for creating software out of necessity. A programmer who writes software can describe the program he has written (the wish) but typically does *not* know and cannot even describe the physical form that the resulting software takes (electrical signals in a computer's memory). If a programmer is to describe his software *at all* in a software patent, the only feasible way to do so in most cases is in terms of the program that was used to create it.

This is not necessarily true for today's artificial inventions. You could directly describe the physical structure of an automobile frame that you create with artificial invention software even though you invented the frame by wishing. But tomorrow's wish patents might still arise of their own steam in another way. Imagine you're still an automotive engineer. You've filed a patent application on the car frame you invented using artificial invention technology. Then you realize you could have done more. When you gave the airflow equation to your software, you told it to show you only the single best car frame it could find, and this is the single car frame you are trying to patent. But you could have told your genie to show you the *thousand* best car frames it could find. In reality, artificial invention software often produces a large number of results that satisfy the input (wish) given to it, just as biological evolution often produces a variety of species adapted to survive in the same environment, and a variety of forms (such as various kinds of wing) for performing the same function. Because you're already seeking patent protection for one car frame, why not try to patent them all?

So you go back to your computer, type in our trusty old airflow equation, and have your artificial invention software spit out designs for 1,000 car frames. Assume for purposes of example that all of these frames are in fact new. You start writing a patent application covering all thousand of them by drawing and describing the shape of each car frame separately, just as if you had invented each frame by hand. If each frame satisfies the legal requirements for patentability, then you should be entitled to patent protection for every last one.

After a while, however, you realize there is a way to consolidate your description of the thousand car frames into a single concise description of how to make and use all of them, followed by a single claim to all of them. After all, the textbook airflow formula—the wish you used to invent the frames—captures, very clearly and very succinctly, a property that all of the frames share. On its own, however, merely describing this formula in the patent application might not adequately explain to someone exactly which frames you are trying to patent.

Luckily for you, there is a straightforward solution to this problem. You just need to get your patent-writing brain out of its 19th-century mind-set. Patent law requires that you write your patent application in a way that explains to the public how to *make* and *use* your invention. Although the traditional way to do this for something like a car frame is to show pictures of it and include text describing how to build it, this is not how *you* invented the frame. You invented it by typing the airflow formula into artificial invention software. So why not just write a very simple patent application that essentially says:

> If you want to make the 1,000 car frames that I've invented, buy artificial invention software and type in this formula: [details of airflow formula]. Tell the software to give you the 1,000 best car frame designs. Then pass the designs on to an automobile manufacturing plant, where they can surely build any of the frames you want.

You would probably need to furnish some additional details, such as the kind and version of the artificial invention software that you used (e.g., "Invention Genie® version 2.3"), to make it truly possible for the public to reproduce your results. But the resulting description would be sufficient to enable the public to make and use your invention, and surely it would be more concise than describing a thousand frames separately.

At the end of the patent application, you would write a single patent claim: "I claim any automobile frame produced by inputting the following airflow equation to artificial invention software." Again, this is much easier for you than writing a thousand separate claims.

Although you are on the right track, this particular patent application should be rejected because you have attempted to claim the airflow equation "in the abstract," and the particular airflow equation that you used is not new (you found it in a textbook). Furthermore, you have attempted to obtain patent protection for *all* car frames that artificial invention software can produce using the airflow equation, and some of those frames may not be new. If any such frame already exists, your patent will be rejected, and rightfully so.

But we are on to something here. By attempting to cut down on the amount of writing you need to do in submitting a patent application for a large number of *artificial inventions* (car frames, in your case), you've ended up in effect writing a patent application for a *wish* for those artificial inventions. The single claim of that patent, if it is allowed, would potentially cover *any* car frame that could be produced by this wish using an artificial genie.

Therefore, such a patent would not technically be a patent on the wish itself, in the sense that the patent—if interpreted correctly—would not empower you to stop other people from writing the wish down, discussing it with each other, learning from it, or even improving on it. Rather, writing the patent application *in terms of the wish* is a shorthand way of patenting all of the car frames produced by the wish once the wish is turned over to an artificial genie. The patent application is still an application for a patent on the resulting inventions (car frames), just written more abstractly by succinctly describing a feature that characterizes the entire *class* of inventions rather than describing each instance of the class individually.

In one sense, writing patent applications in this way is nothing new to patent law. If you invent a chair whose novel feature is a new geometric configuration of legs that affords increased structural support over previous chairs, you can write a patent application for the chair by describing the geometric configuration of the legs, showing an example of the chair, and—assuming you've satisfied all of the other requirements for patentability—the resulting patent will cover *all* chairs having legs arranged in that configuration, whether they are made of wood, metal, or plastic; whether the back of the chair is solid or slatted; and whether the seat is flat or concave. The patent, in other words, covers the entire *class* of chairs having the novel feature that you invented. Much of what other patent lawyers and I do for our clients is to assist them in identifying the broadest possible conception of their invention and capturing this broadest conception in a patent application in an attempt to make the application cover the *class* of objects falling within that conception, rather than just a single object.

Returning to your car frame patent application, although we started by asking whether individual car frame designs (the *output* produced by a genie) are patentable, we then rewrote your patent application in terms of the wish (the *input* provided to the genie). Put yourself in the position of the patent examiner who reads your patent application, which describes the wish (and possibly the particular genie) that you used to produce car frames, and concludes with a claim on the wish. The examiner will, in effect, need to determine whether the wish is patentable, even if in the end the patent application is one for car frames rather than a wish.

In Search of Patentable Wishes

Although your first stab at a wish patent application will be rejected because your wish is not new (you got it from a textbook), it is not hard to imagine cases in which a wish could be both novel and nonobvious. This doesn't mean it would be easy to write such a wish. Gillette may have thought it had conceived of a broad new class of razors when it introduced a razor with five blades,[3] but the satirical newspaper *The Onion* had prophetically lampooned the idea a year before the Gillette razor was announced.[4]

Although difficult, it is far from impossible to design novel and nonobvious wishes. Every valid software patent is proof of this. If you write a computer program that describes a new and useful antilock braking algorithm, you can process that program using multiple compilers to automatically create multiple versions of the "same" software, all of which are described by the computer program you wrote. (This is multiple realizability in action.) Every time a programmer creates novel and nonobvious software, she has written a novel and nonobvious wish.

Similarly, a significant amount of work being done in the field of genetic algorithms today involves designing new and better fitness functions, which are examples of wishes. Furthermore, we can find examples of novel and nonobvious wishes far back in the history of patent law. Consider claim 8 of Samuel Morse's telegraph patent, which claimed not only the telegraph itself but more broadly:

> the use of the motive power of the electric or galvanic current, which I [Morse] call electro-magnetism, however developed for marking or printing intelligible characters, signs, or letters, at any distances, being a new application of that power of which I claim to be the first inventor or discoverer.[5]

In lay terms, Morse claimed that he was entitled to patent protection not merely for the particular telegraph he had described in his patent, but more generally for *any* device that could transmit and record electronic messages over arbitrarily long distances—just as you are attempting to claim patent protection for *any* car frame consistent with the airflow formula that you used as a wish. The key difference between the airflow formula and Morse's claim is that, before Morse's invention of the telegraph, no device actually existed that satisfied his wish. His claim, in other words, described a novel, nonobvious, and useful wish.

You might expect, then, that Morse was entitled to a patent on all forms of devices for transmitting messages using electricity, particularly because he did describe in his patent how to make and use his wish-come-true: the telegraph. The U.S. Supreme Court, however, *struck down* Morse's claim in 1854 because although his patent sufficiently described how to make and use the *telegraph*, it did not sufficiently describe how to make and use what he claimed: *any* device for transmitting messages. The Court nullified Morse's patent claim because he had tried to grab too much territory with it.

It would be a mistake to interpret the *Morse* case to stand for the proposition that wish patents should never be granted. Rather, I interpret *Morse* to mean that you are not entitled to a wish patent *if your patent does not enable a person having ordinary skill in the art to make and use a broad range of inventions falling within the scope of your wish.* If you try to patent an airflow formula as a wish for making car frames, you had better explain in your patent application how someone could use that airflow formula as an input to an artificial genie to produce *any* car frame that satisfies the formula. If you fail to do so, your wish patent should be rejected.[6]

We have good reason to believe, however, that inventors in the Artificial Invention Age will increasingly succeed at obtaining wish patents where Morse failed. Because prominent varieties of artificial invention technology excel at *searching* for inventions that satisfy a given wish, such technology often produces many possible solutions to a given problem—and will keep trying to generate such solutions until you tell it to stop. Although an artificial invention genie may hit bumps in the road or even stop bearing fruit at some point, as artificial invention technology improves and computing power increases we should expect artificial invention genies only to improve at fleshing out a broad range of possible solutions to a given problem. If you have such a genie at your disposal, you will be able to write a patent for a wish that can describe a variety of wishes-come-true, thereby increasing the likelihood that a broad patent claim to the wish itself will be upheld where Morse's failed.[7, 8]

The Case for Wish Patents

Now that we've seen how wish patents could arise naturally out of the interaction between artificial invention technology and extant patent law, I will go one step further and argue that allowing patents on new, useful, and nonobvious wishes promotes innovation. In other words, *wishes should be patentable*. There, I've said it.

Just as every culture has its own creation myth, so too do legal theorists tell stories about how patent law arose as a mechanism for promoting innovation.[9] Such stories have venerable roots in the work of Hobbes, Rousseau, Locke, and others who described how humans were created in a "state of nature" lacking rules and then developed social norms and eventually laws to govern their conduct. Such tales of the origins of law need not be interpreted as accurate historical accounts; they are valuable as "rational reconstructions" explaining why certain laws are socially beneficial, independent of the complicating details of particular existing legal systems.

Consider, then, just such a story about why we should grant patents on wishes, a story closely paralleling modern justifications for patent law generally. Imagine a world in which everyone had a true genie, not an artificial one, at his or her disposal. If you wanted a cup of coffee in such a world, you would just need to say out loud, "Genie: coffee, please," and a hot cup of freshly brewed java would appear in your hand. If you were more imaginative, you might wish for objects that did not yet exist ("Genie: make me a thought-controlled car." "Genie: replace my television with one that displays holograms." "Genie: produce a pill that cures all forms of cancer.").

You might think that patent law would be not only unnecessary but positively harmful in such a world. If you instructed your genie to produce cancer-curing pills and then obtained a patent on the pills, you would have the legal right to stop anyone else from making or using the same pills. Yet such a patent would be unwarranted if anyone else easily could have used her own personal genie to produce the same pills merely by wishing for them.

Even more troubling is a scenario in which you obtain a patent not simply on the pills themselves but on the wish that produced them. Such a patent would enable you to block people from making the same pills that you made and from making *any* pills that a genie produces when instructed to "produce a pill that cures all forms of cancer."

This cautionary tale does not imply, however, that wishes should be *per se* unpatentable. Why? Wouldn't people, equipped with powerful genies, inevitably produce every invention we could ever need? Not necessarily.

Think about every story of genies or wishing you've ever heard. Does the person in the shoes of Aladdin end up healthy, happy, and wealthy? Inevitably not. Instead, he ends up sick, miserable, and in financial ruin. Why? In part because greed clouds judgment. But even more important, would-be Aladdins find that formulating wishes is not as easy as it first seems. Just remember King Midas, who wished that everything he touched would turn to gold and then found himself cursed, unable to eat as each morsel in his hand turned to gilded stone.

Poor King Midas at least had the benefit of starting with a clear idea of his goal: to be rich. Most of us have as much trouble clarifying our desires as expressing them precisely in words. Being a successful wisher requires both of these skills.

In a world of genies, therefore, people's wishes would probably fail frequently, often backfiring on their wishers in harmful and unexpected ways. Even when wishes were successful, they might only privately benefit the wisher, not society as a whole. You might formulate an unusual wish for a new kind of rocket engine but decide to keep the wish a secret, depriving the public of its benefits. If everyone in the world wished for a personal stockpile of gold, the planet might be overrun with mounds of bullion. Unbridled, and uncoordinated, wishing wouldn't necessarily maximize public welfare, even with the most powerful of genies at our beck and call.

If people failed miserably and repeatedly enough as wishers, they might become so frustrated with the process that they would stop wishing entirely. This outcome would also not benefit society as a whole; we would lose out on all of the potential benefits of genies.

More likely is a third scenario, in which most people would be horrible wishers, while a few would develop particularly strong wishing skills. They would learn how to clarify their goals and how to express their wishes in a precise language that would reduce the likelihood of their wishes backfiring. They might even develop a common "wishing language" that could be used to make a variety of wishes successfully.

Skill at wishing would be valuable in such a world. If you had wished once and been burned, you might try hiring a professional wisher for your next wish. Want the perfect body? Instead of simply asking your genie for a "perfect body" and risk being turned into a statue or a perfect frog, just call 1-800-WISH-PRO, where a professional wisher is standing by to interview you about your goals and then formulate a wish designed to achieve it—for a fee. She would write the wish down on paper for you, and you would read it aloud to your genie to make it come true.

Markets would develop in wish writing, with the rates charged by professional wishers determined by supply and demand. Not every wish, however, would need to be a custom job. Large numbers of people might desire the perfect body but be unable to afford the cost of hiring a professional wish writer to achieve their goal. In this case, a wish writer might write the "perfect body" wish once and then sell the wish itself (on paper, or for download over the Internet) on the mass market for a lower price made possible by spreading the cost of writing the wish across a large number of customers. Markets would develop in wishes themselves, with prices again being set by supply and demand.

Where there are markets, there can be market failures. Imagine that it cost our professional wish writer $10,000 to write the "perfect body" wish, and that she began to sell copies of the wish to the public at $10 each. She therefore would need to sell at least 1,000 copies of the wish to make back her initial return on investment. What is to stop her first customer, however, from purchasing a copy of the wish and then reselling it to others for $5 per copy, thereby undercutting the wish's author? In such a scenario, it might be impossible for the wish writer to set prices at a level that would enable her to turn a profit. In response, she might stop writing wishes and go back to her day job, or revert to writing custom wishes for the wealthy at $20,000 each.

The public again loses in this scenario, because wishes would not be widely available at low cost. This, however, is exactly the kind of breakdown that traditional patent and copyright law is intended to remedy. If the cost of *copying* a work, whether it be a poem, invention, or wish, is relatively low compared to the initial cost of *creating* it (sometimes called "the cost of making the first copy"), then an unregulated market may be inadequate to produce the optimal public benefit from production of the work. In this case, granting the creator (whether author, inventor, or wisher) the ability to exclude others from making certain uses of the work can restore the balance the unregulated market could not maintain. Applied to the world of genies, permitting wish writers some ability to leverage the law to stop others from copying their wishes without authorization could encourage wish writers both to keep writing new wishes and to make those wishes available to the public. The same logic applies to the products (such as cancer-curing pills) the wishes produce.

Lest you think that all this talk of buying and selling wishes, and of using copyright and patent law to protect them, is pure fantasy, the fairy tale I've just told closely parallels the actual history of copyright and patent protection for computer programs—the most primitive kind of "wish" in the Artificial Invention Age.[10] Furthermore, the arguments I've suggested for and against allowing patents on wishes closely parallel the debates that have raged over software patents in academic journal articles,[11] court decisions, and national legislatures for decades.[12] The challenge both fictional wishes and real-world computer programs pose for patent law is that on the one hand neither of them appears to be patentable because both are a kind of *instructions* describing a result that the wisher/programmer wants to achieve, and traditionally instructions are not patentable. On the other hand, the existence of genies in the fictional case and computers in the real case make it possible for wishers and programmers respectively to cause their instructions to be carried out *automatically*—without any further manual effort—to produce results that traditionally *are* patentable, such as a toothbrush or software for controlling the brakes of a car.

As a result, both fictional wishes and real-world computer programs can be traded in the marketplace as *substitutes* for the end products they can be used to generate. As a consumer, you might place the same value on a bottle of cancer-curing pills as you do on a wish for such pills, assuming that you have access to a genie that can transform the wish into the pills themselves at no additional cost. If the wish is the functional equivalent of the end product, then it is at least plausible to apply the same legal regime—patent law—to the wish as to the end product itself.[13]

This simple story reflects the primary motivation for copyright and patent law: to encourage people to create works, such as novels, poems, machines, industrial processes, and (now) wishes, and to make those works available to the public. If, however, we let market mechanisms alone govern the production and distribution of creative works that are *expensive to produce yet inexpensive to copy*, we may produce a system that either (1) underproduces such works or (2) encourages authors and inventors to make their works available only to a select few, rather than to the general public. Such "market failures" are precisely the situation copyright and patent law are intended to address.[14]

Wishes are prime candidates for patent protection, therefore, because they can be useful, difficult to write, and easy to copy. Furthermore, as artificial invention technology spreads, people are likely to develop more specific and advanced skills in wish writing (such as designing better fitness functions for genetic algorithms), and wish writing is likely to branch off from the design of artificial invention technology itself into its own field, just as computer programming branched off from electrical engineering in the 1960s. If the history of the computer industry is any guide,[15] we could see artificial wishes being bought and sold on the open market as discrete products, with wish writing *surpassing* artificial invention technology development in its economic value. As artificial invention technology enables wishes written at higher and higher levels of abstraction to accrue distinct economic value, we should ensure that patent law remains flexible enough to follow the value upward in response.

Interpreting Wish Patents

Even though wishes should be patentable, we cannot simply apply existing patent law as is to wishes, for a variety of reasons. For example, a patent on a wish will by default cover *any* product that can be produced using the wish. We should allow wishes to receive such broad legal protection only if doing so strikes the right balance between offering an incentive to the wish writer and leaving sufficient room open to the public to solve the same problem as the wish in other ways.

For example, a patent on a new, useful, and nonobvious airflow formula should not give its owner the power to stop anyone else from thinking about the formula or writing it down on paper. Protecting against such perverse outcomes will require patent offices and courts to *interpret* wish patents narrowly to cover only the products the wishes produce, and possibly the process of using particular wishes to produce those products. This is essentially the status quo in the context of software patents.

Such questions about how we should *interpret* wish patents point out another potential problem with them. Because our wish patent for a car frame describes the wish in terms of an abstract airflow equation, the patent may not present the public with a very *clear* indication of which specific car frames the patent does and does not cover. Failure to ensure such clarity would cause the patent to violate the "public notice function" of patents.[16] One of your competitors should be able to design a car frame that competes with yours and use your patent to let him know whether his car frame design infringes your patent or not. If your car frame patent is too vague, then competitors may decide it isn't worthwhile for them to take the risk of designing alternate car frames and facing a patent infringement suit from you. As a result, the public will be deprived of such additional products.[17]

Been There, Done That: Interpreting Software Patents

For evidence that wish patents may be susceptible to being both particularly broad and vague (unclear), we need look no further than the history of software patents. James Bessen and Michael Meurer of the Boston University School of Law argue that although we have always needed to guard against fuzzy patents, the U.S. patent system is producing increasingly fuzzy patents,[18] and that software patents are particularly susceptible to creating fuzzy boundaries and therefore warrant special attention.[19] More generally, the history of software patents has been riddled with controversy over whether software patents are too *abstract* or too *vague*.[20] If the battle over patents on computer programs—Wishes Version 1.0—is any indication, we are in for an all-out war over even *more*-abstract wish patents.

The furor over software patents is unique in the history of patent law, which has accommodated previous revolutionary technologies, from the internal combustion engine to the light bulb to the transistor, without so much as a hiccup. Furthermore, in other instances where new types of patent have sparked debate, the reasons for such debate are clear. Patents on pharmaceuticals and gene sequences are controversial because they raise ethical questions about the ownership of life itself and the affordability of health care, two issues that don't apply to software. Controversy over patents on auctions and other kinds of "business methods" can be understood in light of the fact that business methods historically have not been considered a kind of technology—a criticism that does not apply to computer software.

Furthermore, as Besson and Meurer have pointed out, in all other instances where the public has called for a ban on patents for a particular kind of technology, the industry *behind* the technology has stood as the strongest defender of such patents.[21] In the case of software, however, giants in the industry, including Microsoft, Oracle, and Autodesk, testified before Congress and the Patent Office in *opposition* to software patents as recently as the 1990s, arguing that expanding patent law to encompass software would harm the software industry and that innovation in software was proceeding at blinding speed without software patents, thank you very much.[22]

You might think that when Microsoft speaks, the legal system would listen and simply not allow patents on software. Yet the U.S. courts have given increasingly wide berth to software patents, and Congress has taken only minor steps to rein them in. The basic reason no one has been able to stop the forward march of software patents is that, for the reasons given earlier in this chapter, a software patent application is merely a shorthand, abstract way of describing how you can use a computer program to reconfigure a computer—an indisputably physical machine—to do what you want it to do. If a computer that has been reconfigured *manually* to control the brakes of a car is a patentable improvement to an existing computer, then there is no clear basis for prohibiting a patent on the same reconfigured computer merely because it was reconfigured *automatically* by programming it.

Any attempt to impose an outright ban on software patents would simply lead crafty patent lawyers to rewrite their software patent applications as longer, more detailed applications describing a variety of physical computer configurations that you could use to implement any particular piece of software, just as a ban on wish patents would lead you to write your car frame patent as a long, detailed patent application describing the thousand car frames that could be produced using your wish. In other words, such a ban would not actually *prevent* software patents from being obtained; it would simply cause people to change the form in which they *write* such applications. Such long, detailed patents would not necessarily be any easier for the public to understand than a concise but abstract wish patent. In fact, the exact opposite may be true; a patent describing a new airflow equation as a wish may offer the public a clearer, simpler idea of the gist of what your patent covers than a convoluted document describing a thousand car frames.

We have experienced exactly this cat-and-mouse game between the legal system and patent applicants in the history of software patents. Patent offices and courts worldwide began to view software patents with a critical eye as long ago as the 1960s, often rejecting such patents as too broad and abstract. For example, the U.S. Supreme Court first wrestled with software patentability in the early 1970s in the case of *Gottschalk* v. *Benson*,[23] which involved software for converting numbers from one form into another. The Court in *Benson* invalidated the patent, holding that to allow it would "in practical effect" be to allow a patent on an "idea," and abstract ideas are not patentable—period.

In response to such rejection of software patents, patent lawyers continued to pursue patents on software but began writing them using words and images that focused on the concrete, physical hardware that was programmed with and used to execute the software, thereby giving such patents a more physical feel. If patent offices and courts were to reject such patents, they would have to do so for the reason that the *functionality* the patents described was not new, useful, or nonobvious, *not* that the patents failed to describe a physical machine having such functionality. This strategy largely worked, at least in the United States, leading the Patent Office and courts to look more favorably on patents that were effectively on software. In more recent years we have come full circle, with the courts in particular becoming more accepting of software patent applications that merely describe the software being claimed, but not the details of the hardware necessary to implement it. The road that led to this equilibrium, however, is littered with much conceptual debris that has yet to be fully cleared.

Software patents have come under even stronger attack from *outside* the legal profession. Software patent critics frequently argue that traditionally patentable machines are concrete and tangible, while software is in some sense abstract and intangible and should therefore not be patentable. John Perry Barlow put this argument in its purest form in his "Declaration of the Independence of Cyberspace," in which he stated, "Your legal concepts of property, expression, identity, movement, and context do not apply to us. They are all based on matter, and there is no matter here."[24] Barlow later expanded on this concept when he claimed that "all the goods of the Information Age [including software] . . . will exist either as pure thought or something very much like thought: voltage conditions darting around the Net at the speed of light, in conditions that one might behold in effect, as glowing pixels or transmitted sounds, but never touch or claim to 'own' in the old sense of the word."[25] Similarly, science writer James Gleick argued against software patents by claiming that "patents began in a world of machines and chemical processes—a substantial, tangible, nuts-and-bolts world—but now they have spread across a crucial boundary, into the realm of thought and abstraction."[26]

Examples of purportedly overly broad software patents abound. In 1992 Compton's Multimedia was granted a patent that it claimed bestowed patent rights on all multimedia, such as any CD-ROM that combines text and graphics together.[27] In 1998 British Telecom began claiming that one of its patents covered all hyperlinking—the fundamental mechanism for following links on the Internet.[28]

The public outcry over software patents has risen to a fever pitch in recent years. When the European Parliament attempted to pass the "Software Patent Directive," opponents took to the streets in protest—not the kind of passion you would expect an otherwise arcane area of law to arouse. Organizations such as the Foundation for a Free Information Infrastructure (FFII) mounted organized opposition to the directive, garnering support primarily from small- and medium-sized enterprises.[29] Richard Stallman, founder of the "free software movement," has gone so far as to argue that it is *unethical* for software to have owners because doing so deprives software users of their natural freedom to use software in any way they please.

Keeping Wish Patents on a Tight Leash

Although arguments that software patents have the *potential* to be overly broad and vague have merit, they do not justify banning them entirely, for at least three reasons. I've already pointed out the first two. First, a ban on software patents would merely drive them underground by causing patent lawyers to write such patents as hardware patents. Second, even the world's most powerful artificial genie won't produce anything without a skilled wisher commanding it, and patent law has a role to play in encouraging people to develop new and useful wishes and make those wishes available to the public.

Third, recall that a wish expressed *abstractly* is not necessarily *vague*. Some statements are both abstract and vague, such as "cats are beautiful," which is abstract because it refers to *all* cats, and vague because beauty is in the eye of the beholder. "The set of Toyota automobiles built in 1995" is abstract but *not* vague because it refers very precisely to a large class of individual instances. Therefore we should not rule out patents on software or wishes *per se* merely because they would be abstract. Our challenge is to make sure that they are not *more* abstract than is warranted and that precision is retained as abstraction increases.

Patent law already includes a host of mechanisms that can be used to keep the potential breadth and vagueness of wish patents in check. The most significant is the *novelty* requirement. Imagine that you submit a patent for your *old* airflow equation, in an attempt to obtain patent protection for *any* car frame—old or new—that can be produced using the equation. This claim should be rejected for lack of novelty if even *a single car frame exists that falls within the scope of the wish.*[30] For example, if the patent examiner can find such a car frame, he should be allowed to reject your entire wish claim for lack of novelty. If you seek a patent on a wish that abstractly describes a *problem* to be solved, then the wish should be deemed novel only if *no* previous solution to the problem exists. This is a high hurdle to surmount and, *if applied rigorously*, will address most of the concerns about overly abstract wish claims.

If your broad wish claim is rejected for lack of novelty, you can attempt to modify your claim by making it *narrower*, amending the language of the claim to carve out preexisting car frames, so that the modified claim covers only using your wish to produce truly novel and nonobvious car frames. Then the patent examiner could again use any existing car frames that overlap with your new, narrower claim to knock that claim out as well, forcing you to make your claim yet narrower again or abandon it. Novelty, and nonobviousness, will play an important role in keeping the breadth of wish claims in check in the Artificial Invention Age.

Of course, the patent examiner doesn't need to go down to the level of individual car frames to knock out your claim. If you claim a wish and the examiner can find evidence that *the wish itself* existed before you claimed to invent it, then your wish should be rejected because it isn't novel even on its own terms. In some cases, you might still be able to patent certain car frames that you've produced using the wish, but not the wish itself.[31] Similarly, even if the examiner cannot find evidence that your wish is old, your wish might be *obvious* in light of previous wishes because, for example, your airflow equation is just a trivial modification to similar equations that automotive engineers have long known to be useful in designing automobile frames. In other words, the examiner should be empowered to examine *your wish in its own terms* for satisfaction of the novelty, nonobviousness, and utility requirements. This is essentially how software patent claims are examined today.

If you are able to surmount all of these barriers, however, then you should be entitled to a patent on your wish, no matter how abstract. We call such patents "pioneer patents," and patent law gives them special treatment by interpreting them particularly broadly.[32] After all, if you've proven that you have invented a wish that is truly capable of producing a wide range of new and useful devices, then patent law should extend to you protection for that wish for all of the reasons we've encountered. Although such a patent may be broad and abstract, it is not *too* broad and abstract, because its far reach is justified by the facts supporting it.

Consider the so-called RSA encryption patent, which is famous among patent lawyers and infamous among RSA's competitors.[33] "RSA" stands for Rivest, Shamir, and Adleman, the three computer scientist-mathematicians who created the RSA encryption algorithm, which scrambles a message so that no one except its sender and intended recipient can read it (at least without gargantuan effort). The RSA encryption algorithm can be used to maintain the secrecy of email communications, financial transactions, medical data, and any other kind of information that can be encoded digitally.

Although many encryption algorithms existed before RSA's, no one then or now questioned that the algorithm was new, even groundbreaking, and extremely useful for encryption. The algorithm, like many pioneering inventions, solved a problem that previously had eluded everyone in the field but was easy to understand once it was known.[34] Best of all, it could be described relatively succinctly and abstractly in the language of mathematics and computer science. As a result, the algorithm could be implemented easily in software on any computer in the world, or directly in circuitry or just about any other kind of device. A written description of the algorithm was a very broad and abstract wish for an encryption machine.

Rivest, Shamir, and Adelman obtained a patent on the algorithm, and a broad one at that. The patent was written by savvy patent lawyers who understood how to make it cover essentially any method of implementing the algorithm, whether in software, hardware, or any other kind of device. It did just that for 20 years, raking in untold millions in royalties for RSA Corporation; anyone who wanted to use the RSA encryption algorithm for any purpose on any kind of device was required to pay for the privilege of doing so. RSA's competitors reportedly threw parties around the world on the day the patent expired.[35]

The RSA encryption patent is a close modern analog to Morse's patent claim 8 in the sense that it covered essentially any device that could transmit messages encrypted using the RSA algorithm, just as Morse's claim 8 covered essentially any device that could transmit messages electronically. The RSA algorithm was described and claimed succinctly and abstractly in the RSA patent, just like Morse's claim 8. Yet RSA succeeded where Morse failed, in large part because the advent and widespread adoption of general purpose programmable computer technology made RSA's abstract description of their algorithm capable of enabling anyone to implement it in a device using only ordinary skill. In other words, RSA succeeded where Morse failed because they invented the RSA encryption algorithm in a world populated by artificial genies.

If Rivest, Shamir, and Adelman had devised their algorithm on paper a century ago and included a description of it in purely mathematical terms in a patent application, such an application would likely have been rejected—and rightly so, because such an abstract description would not have enabled a person of ordinary skill in the fields of mechanical or electrical engineering to build a working device that could carry out the algorithm. Why? Because general purpose programmable computers, capable of transforming abstract mathematical descriptions into reality, did not exist. Therefore if you wanted to build an RSA encryption *machine* at that time based merely on a description of the *algorithm*, you would have had to *invent* such a machine, by which I mean using something more than routine skill to design physical components to carry out the algorithm. Translating the abstract written description of the RSA encryption algorithm into a working machine—granting the wish—would have required exactly the kind of *unordinary*, inventive, skill for which patent protection is reserved. This is exactly the set of real-world conditions under which a patent application that merely described the RSA encryption algorithm in abstract terms would not "enable a person having ordinary skill in the art to make and use the invention . . . without undue experimentation."[36] Such a patent application would therefore be rejected. Yet this same description became patentable in the late 20th century because general purpose computers had become widely available that could transform the description automatically into working software without *any* additional human effort.

As the RSA case demonstrates, abstract patent claims are not necessarily problematic so long as their broad scope is justified by the breadth of the underlying invention. The legal system, however, must examine patent applications for wishes particularly carefully to ensure that abstract claims that are *not* justified by the evidence do not slip through the cracks.

Potential Pitfalls

Because the legal standards I've suggested here require reinterpreting existing legal rules rather than creating entirely new rules from scratch, these standards would best be implemented primarily by judges, patent office officials, and patent lawyers in the course of handling individual patent applications, rather than by legislative changes to the patent laws themselves. Applying such case-by-case adjudication to software patents, however, has produced a hodge-podge of inconsistent rulings over several decades that we have yet to fully sort out.[37] We can't afford to wait another 30 years while the patent system adapts to artificial invention technology; the technology is developing too quickly and the economic stakes are too high. To obtain the benefits of case-by-case adjudication while minimizing its shortcomings, those responsible for administering the patent system in the Artificial Invention Age should be aware of the problems they will face so that they can be prepared to tackle them coherently in individual cases.

Distinguishing Ideas from Applications

The abstractness of wish patents will make those patents abut the dividing line between unpatentable abstract ideas and patentable practical applications of abstract ideas more closely. As a result, it will be more important than ever to pay attention to whether a wish, as it is described and claimed in a patent application, is merely an abstract idea or a practical application of one. Consider just a few historical examples of cases in which courts have had to draw a line between the two.

- In *Neilson* v. *Harford*,[38] the defendant argued that Neilson's patent on a blast furnace was in effect a patent on the abstract scientific principle that hot air feeds a fire better than cold air. The court disagreed and upheld the patent on the grounds that it claimed not a principle but a *specific machine*—the blast furnace—embodying the principle, and that the inventor had described in sufficient detail how to make and use that machine.
- In *Le Roy* v. *Tatham*,[39] the defendants argued that Le Roy's patent on an improved lead-pipe-making machine was in effect a patent on the scientific principle that lead holds its shape when cooled. The U.S. Supreme Court agreed and invalidated the patent.

- In *Mackay Radio & Telegraph Co.* v. *Radio Corporation of America*,[40] the U.S. Supreme Court upheld a patent on a V-shaped antenna in which the prongs of the V formed an angle that had never been used before in such an antenna, even though the angle was derived from an equation that had been well known for years and was not even developed by the antenna's inventor. The Court deemed the antenna to be patentable because although "a scientific truth, or the mathematical expression of it, is not patentable invention, a novel and useful structure created with the aid of knowledge of scientific truth may be," and the V-shaped antenna was such a structure.

If you find it difficult to reconcile the decisions in these cases with each other, join the club. These and many other cases demonstrate the difficulty of distinguishing between abstract ideas (such as laws of nature, mathematical formulas, and scientific principles) and practical applications of those ideas. This dividing line, however, is the best we have, and there is good reason for continuing to uphold it. The U.S. Constitution says that the purpose of patent law is to promote the progress of the useful arts, and abstract ideas do not have any immediate practical use. It is only when someone has made it possible to incorporate an abstract idea into a particular physical object or process having a direct practical benefit—a practical application of the idea—that the idea develops the kind of "practical utility" that patent law is intended to protect.

Although it has always been difficult to distinguish between abstract ideas and practical applications, patent law did a pretty good job at keeping the dividing line clean for most of its history. In the end, it was easy enough to tell the difference between a blast furnace and a law of nature. Try standing very close to the furnace if you question whether it is an abstract idea.

Software changed all this. One reason software patents raise the abstract/practical distinction more frequently than other kinds of patents is that inventions in software patents tend to be described using abstract *words* that can be interpreted to refer either to physical entities or to abstract ideas. Take the word *data*, which can refer to something physical, such as ink representing numbers on a page, or to an abstract idea, such as a thought in your mind or a pure number having no physical form. As a result, when such words as *data, record, field,* and *bit* appear in software patents (as they often do), it isn't clear whether we should interpret them narrowly to refer only to the *physical incarnations* (sometimes called "instantiations" or "embodiments") of those entities or more broadly to include abstractions.[41] We don't have this problem with traditional patents on car frames, because the terms used to describe and claim a car frame, such as *steel* and *fender*, refer unambiguously to physical entities.

The easy solution to this problem is to require the *inventor* to specify very clearly and narrowly what he is trying to patent, so as to leave no room for confusion. If an inventor thinks he has come up with a patentable way to reorganize data on a hard disk drive, then he should say in his patent application that he has invented "a method for reorganizing data *stored in tangible signals on a hard disk drive*," rather than that he has invented a mere "method for reorganizing data," leaving us in the lurch as to whether he is trying to patent such a method in the abstract or as embodied in a machine. If the inventor fails to be specific enough, then his patent application would be rejected.

Another option is for the Patent Office and courts to interpret abstract patent claims *as if* they were written more narrowly, by interpreting a phrase such as "a method for reorganizing data" to mean "a method for reorganizing data stored in some physical form" and then analyzing it as such. Either solution should rule out the possibility of granting patents on thoughts or abstract ideas.

We need to be careful, however, not to require inventors to describe and claim their inventions at *too low* a level of detail. When a programmer writes computer source code and then compiles it to create working software, the programmer may not even *know* what the logical or physical structure of the resulting software is and therefore may not be able to describe it in anything other than abstract terms. This is analogous to a chemist who combines chemicals A and B to produce a new chemical C. She might perform experiments on chemical C demonstrating that it can be used to treat skin rashes even though she doesn't know what the chemical structure of chemical C is and therefore cannot describe chemical C in any way other than as "the chemical you obtain when you combine chemicals A and B."[42] Patent law typically allows such an inventor to describe her invention in this more abstract way when supplying a structural description is impossible; the same rule should apply to software and wishes.[43]

Hardwiring Manual Invention into the Law

A more difficult problem is that certain 19th-century ways of thinking about inventing have been "hardcoded" into patent law in ways that no patent examiner or court can ignore. In the Industrial Age, inventors typically had no choice but to engage in structural design because they did not have computer genies at their disposal to do it for them. As a general rule, patent law tracked this real-world need by requiring inventors to describe and claim their inventions in terms of physical structure. Imagine, for example, that a patent included only a *functional* description of an invention, by which I mean a description of what the invention *does* rather than *how* the invention does it. If Morse had described the telegraph only *functionally*, by describing it as "a machine for receiving a sequence of taps from a person, transforming those taps into electrical signals, and transmitting those signals on a wire over arbitrary distances," such a purely *functional* description would have left out the critical *structural* details necessary to enable anyone to make and use a working telegraph, and his patent application would rightfully have been rejected.

Patent law has always been skeptical of patents that describe inventions in purely functional terms for another reason as well. If Morse had described the telegraph only in functional terms, the public would have been justified in questioning whether Morse actually had invented any such device. Perhaps he was trying to pull a fast one on the Patent Office. After all, if he had truly invented such a machine, why did he not describe its physical structure and show drawings of it in his patent application?

U.S. law goes so far as to hardwire a bias against functional language into the Patent Act itself, by requiring that if an inventor writes a patent claim in *functional* language, the claim must be interpreted to cover only the specific *physical structures* described in the patent's specification.[44] For example, if a present-day Morse were to write a patent claim for "a machine for receiving a sequence of taps from a person, transforming those taps into electrical signals, and transmitting those signals on a wire over arbitrary distances," the claim would *not* be interpreted to cover *any* possible device fitting that description, but rather to cover *only* those specific structures that the inventor had described how to make and use in the main body of the patent application. This section of the U.S. Patent Act says, in other words, that if you try to move upward across the line between Functional Design and Structural Design in Figure 7 (Chapter 5), the law will whack your protection back down to the Structural Design level.

This kind of skepticism may have been warranted in the Industrial Age, when writing a purely functional description of an invention usually was not sufficient to enable anyone to make and use it. But times have changed. Computers have automated the process of transforming functional descriptions—

in the form of computer programs—into working machines. The original factual basis for the antifunctional bias that has been hardcoded into the U.S. Patent Act no longer exists.

The Patent Office and courts have no flexibility to undo this section of the statute, or to ignore it. They have authority only to interpret the law as it is written. Therefore this section of the patent law ties the hands of the Patent Office and courts into interpreting the law as if technology were frozen in the 19th century. This can produce only perverse results. Therefore I recommend that we scrap this section of the Patent Act and instead interpret claims rigorously on a case-by-case basis, using the normal rules of claim interpretation.

The Flood to Come

If I'm right that artificial invention technology will make it possible for inventors to obtain more and more sweeping patent claims, then savvy inventors and high-tech companies will increasingly write their patents with ever broader claims in an attempt to capture as much ground as possible. This is not idle speculation. The phenomenon of patent floods is well documented.[45] Patent floods tend to occur in response to changes in the law that allow a new category of technology to be patented, or to the advent of new fields of technology that serve as a platform for innovation. Both types of causes of patent floods exist in the context of artificial invention technology.

Software was regarded largely as unpatentable by technology companies and patent lawyers into the 1990s. In the early part of that decade, however, a series of court cases eliminated a variety of barriers against patenting software. As a result, in 2002 alone 25,000 software patents were filed in the United States, an eighteenfold increase over the last 20 years. Opening the door to software patents in the United States had a ripple effect worldwide, leading to similar patents in Europe, Japan, and elsewhere. Software giants such as Microsoft and Oracle, which had testified in Congress against allowing software to be patentable as late as 1994, are now leading the field in applying for software patents. Just a decade later, in 2004, Bill Gates issued a memo announcing that Microsoft's new goal was to file 3,000 new patent applications *per year*, almost 13 percent of the total number of software patent applications filed in the country just two years earlier.[46] That's 60 patent applications a week, 12 a day, or 1.5 an hour in an eight-hour workday. The business environment for patents can and does change quickly in response to changes in the law.

An even more recent example is that of business method patents. By one count, the number of financial patents granted in the United States rose by 167 percent year over year in 1998—the year that the Federal Circuit eliminated the prohibition on business method patents. The number of such patents increased again by 65 percent in 1999.[47]

As would be expected, we are already seeing a trickle of patents on artificial invention technology and the inventions produced by it. Stephen Thaler of Imagination Engines has obtained a patent not only on the Creativity Machine itself—a digital genie—but on software created using it. John Koza, the creator of genetic programming, has obtained patents on genetic programming techniques, a patent on various PID and non-PID controller circuits,[48] and on an automated method for designing such controllers—another digital genie.[49] David Goldberg at the University of Illinois has obtained a patent on the hierarchical Bayesian Optimization Algorithm or hBOA (a problem-solving procedure that can optimize designs in engineering systems, among other things).[50] Goldberg's website also announces that the algorithm can be licensed, presumably for a reasonable fee, from the University of Illinois, and that "because of its power and speed, hBOA will provide its licensees with a qualitative and quantitative competitive advantage over their unlicensed competitors."[51]

Let the games begin.

If history is any indicator, we will soon see a flood of patents on artificial invention technology, the inventions it produces, and wishes themselves, just as we have seen in other fields. Although patent floods can be an indicator of a flood of innovation, there is always a risk that the patents themselves will stifle innovation, perhaps because they create a "patent thicket"—a situation in which no one in an industry can innovate without obtaining permission from a large number of other patent holders and because the cost of doing so is prohibitively high.[52]

The historical pattern has been for a patent flood to be followed by after-the-fact attempts to reform the law to rein in the worst impacts of the flood. In the case of the upcoming Artificial Invention patent flood, we have the foresight to see it coming and the opportunity to act now to update the law as necessary to avoid the problems we've experienced in the past. It is particularly important that we do so because artificial invention patents will grant their owners ownership not only over individual inventions but also over the very right to engage in the act of inventing.

9 Free the Genie, Bottle the Wish

SHOULD ARTIFICIAL GENIES, such as the Creativity Machine and future advances in genetic algorithms, be patentable? In this chapter, I recommend that significant advances in artificial invention technology—newer and more powerful genies—should be available for use by the public at little or no cost. Keeping artificial invention technology open, combined with private control over wishes and the new products and processes those wishes produce, will maximize innovation in the Artificial Invention Age. Achieving this result, however, will be challenging and require enforcement of legal rules that may surprise you.

Open Platforms, Closed Applications

Platforms

Forget about the law for now. Instead, let's focus on what maximizes innovation in the real, nonlegal, world. This is a good starting point; any effort to create a legal system for maximizing innovation needs to begin with an accurate understanding of how innovation actually occurs. Only then can legal rules be developed that help to strengthen the forces that increase innovation and weaken the forces that stand in innovation's way.

Every would-be innovator needs tools and raw materials with which to work. Stone Age hunters seeking a better spearhead required a supply of stone and the recognition that sharper was better. Computer programmers need computers, programming languages, and knowledge of how to use both.

Drug researchers demand labs and biology. Authors must have pencil and paper and a language in which to write.

I use the term *platform* to refer to the combination of technologies and knowledge that innovators use as a common base on which to innovate.[1] Another, somewhat narrow, meaning of platform is a combination of basic hardware and software in a computer, such as the combination of Microsoft Windows operating system and Intel microprocessor that form the Wintel platform.[2] Programmers use this platform to write "application" software—such as word processors, Web browsers, and videogames—that can run on any computer having the Wintel platform. One way in which such a platform is a "platform for innovation" is that by providing certain basic functionality that is useful for writing software, such as the ability to store information in memory and transmit information over the Internet, the platform eliminates the need for programmers who write application software to create such functionality themselves. Instead, they can create applications "on top of" the platform by leveraging the features built into the platform. Although a computer platform furnishes features useful for writing software, generally an *innovation* platform may contain components ranging from raw natural resources to manmade machines and processes to languages to fields of knowledge.

I refer to innovations that are built on a platform as *applications* of the platform. If roads are a platform, then automobiles are an application built on the road platform. So too with the electrical grid and household appliances, transistors and transistor radios, and computer hardware and software.

Open Platforms Spur Innovation

You might also see how platforms can be good at spurring innovation in applications built on those platforms. The electrical grid spurred widespread innovation in electrical and electronic devices, the telephone system spurred innovation in products and services that operate over telephone lines, and the Internet continues to spur innovation in Internet-based products and services ranging from email to Web-based retailing to financial services.[3] Although in some cases platforms arise only accidentally, in many cases they are designed with the specific intent of spurring innovation in a particular field. Current examples are cell phone networks and nanotechnology, in which companies and governments are investing significant sums largely in anticipation of future returns from applications that will be developed on the platforms being created today.

The recent history of at least two of these innovation platforms—the Internet and operating systems—gives us good reason to think that platforms promote innovation best when they are "open." This claim is hard to define, much less to prove, and rests on both theoretical and empirical foundations. Others have devoted entire books to defending it,[4] so I won't attempt to do so in great detail here.[5] But some examples should suffice to demonstrate that at the very least we should take steps to ensure patent law does not foreclose those who wish to maintain artificial invention technology as an open innovation platform from doing so.

By an open platform I mean that anyone can innovate on the platform without obtaining permission from someone else to do so.[6] It may not be inexpensive or easy to innovate on an open platform by yourself, but if that is what you choose to do then you can do so and no one has the legal right to stop you. Public roads are an open platform for transportation. Although there are certain requirements for traveling even on public roads (such as a valid driver's license and speed limits), in general you do not need to obtain permission from the government or any private party to travel on them. Not true for traveling through my backyard, or the roads of Disneyworld, which are "proprietary" (private property), and from which you can be excluded by their "owner" for any reason or no reason at all.

Innovation on the Internet is open in the same way. If you have the skill, the desire, and the resources, you can write software that runs on the Internet and just start running it over the Internet, using open Internet protocols, without obtaining permission from anyone. Information about the protocols themselves is available for anyone to access for free.

There is a vast body of literature documenting how initial decisions to make the Internet protocols open contributed to the explosion of innovation on the Internet and its rapid adoption.[7] In addition to low-level protocols such as TCP and IP, many higher-level protocols, such as the Hypertext Transmission Protocol (HTTP) that forms the basis of the World Wide Web and the Simple Mail Transmission Protocol (SMTP) for transmitting email, have also been open from their inception.[8] The success of these protocols and the relative decline of closed systems, such as those used by the once-giant America Online, Prodigy, Compuserve, and Juno, give an indication of the power of openness.

Some use the term *free* instead of open. For example, Richard Stallman of the Free Software Foundation coined the term *free software* to refer to software that, once distributed publicly, cannot be restricted in how it is studied, copied, modified, or distributed. In response to confusion over what is free about free software, and in particular about whether free software can be sold, Stallman has famously stated that you should think of "'free' as in 'free speech,' not 'free beer.'"[9] In other words, what makes free software free is not that it is available at no cost (although it often is), but rather that your liberty (freedom) to use it cannot be taken away by anyone.

Others choose to use the words *open source* instead of free software, both to avoid the baggage associated with the term *free* and for other reasons.[10] I use the term *open platform* in the same way. The fact that a platform is open for innovation does not mean that components of the platform cannot be sold or even patented, but rather that no one has the right to stop you from studying the platform, innovating on it, and distributing your innovations if you choose to do so.[11] Public roads are an open platform for transportation even though a government owns the asphalt and private companies own patents on the composition of the asphalt, because neither of those owners can use those ownership rights to stop you from inventing a new car and driving it on those roads without their permission.

Similarly, a platform is "closed" for innovation if some entity has the legal right to stop others from studying it or producing innovations built on it. The private roads within a gated community or Disneyworld are closed platforms for transportation. Their owners have the legal right to stop you from driving on those roads. Patented inventions are by default "closed" for innovation since no one other than the patent owner can make, use, or sell the invention or certain kinds of modified versions of it without the inventor's permission. Many telephone and cable television networks are closed for innovation in this sense. Individual companies own patents on the technology underlying these networks, which effectively gives those companies the ability to stop anyone else from developing and deploying new technologies that connect to and run over those networks.

You can understand intuitively why open innovation platforms can be particularly good at promoting innovation. Think of mathematics, human languages, and the bodies of scientific knowledge in fields such as physics, biology, and chemistry. All of these are open platforms for innovation. No one would argue that these platforms are free in the sense of it being without cost to innovate on them. Just try setting out to discover a new fact of particle physics and you will learn how expensive it can be to do so. But once someone has made a part of any of these platforms available to the public—a new mathematical formula, word, or scientific fact—anyone else can use it without permission. This promotes innovation for a number of reasons:

- Open platforms facilitate *communication* and *collaboration* among people by providing a shared language.
- Open platforms offer a set of shared building blocks that people can modify and combine to create new innovations.
- Open platforms incorporate previous innovations and make them publicly known, thereby eliminating the need for subsequent innovators to reinvent the wheel and enabling them to focus their energies instead on the cutting-edge problems that remain unsolved. Open platforms are the giants' shoulders on which tomorrow's innovators stand.
- Open platforms reduce costs of entry, thereby reducing barriers to inventors with few resources. This not only increases the pool of potential inventors but works to keep entrenched incumbents on their toes.

These and other features underlie our understanding of why we have kept, and should continue to keep, "basic science" open.[12] They also undergird intellectual property law's historical aversion to allowing language and scientific facts to be copyrighted or patented, as embodied in the prohibition in copyright law on copyrighting ideas and facts,[13] and patent law's prohibition on patenting abstract ideas, laws of nature, and natural phenomena.[14] If scientific principles, once discovered and made publicly available, could be patented, then the public would be unable to make use of objective facts about the world without first obtaining permission from their owner. No one owns English, algebra, or physics, and that is a good thing.

Going past the intuition that certain things should just not be owned, we generally consider the discovery of a new scientific fact, or a new law of nature, to lack the *practical* value that patent law is designed to protect, no matter how much sweat-of-the-brow or even genius went into its discovery.[15] Newton's discovery of the law of gravitation and Einstein's discovery of the law of relativity, no matter how revolutionary, still did not directly enable the public to make and use any particular new machines or processes that were based on those laws.[16] Additional work was still required to use our newfound knowledge of those laws to design new and useful machines based on them. Therefore to have given Newton and Einstein patents on the laws themselves would have been to give them the right to exclude the public from making and using inventions they had not actually taught the public how to make and use. This violates the fundamental *quid pro quo* underlying patent law.

Closed Applications Spur Innovation

Historically we have thought that allowing applications to be closed, such as by allowing them to be patented, is not only acceptable but affirmatively the right way to spur innovation. By an application being closed I mean that it has an owner who has the right to decide, for good reason or no reason at all, to grant or deny permission to others to use that application, whether it be a design for an automobile engine, transistor radio, or computer program. Even those who object vigorously to patents on software often argue that copyright law is the right way for intellectual property law to promote innovation in software, thereby accepting the basic premise that making software applications closed is useful for promoting innovation in software.

Allowing applications to be closed is in fact the basic mechanism that patent law uses to promote innovation. Although I've said all along that patent law is intended to promote *innovation*, patent law grants patents directly only on *inventions*, not *innovations*. Although in colloquial usage these two terms sometimes have the same meaning, they have subtly different meanings— at least to economists and patent lawyers. An *invention* is a new and useful product or process; its distinguishing characteristic is some technical feature that makes it different from all that came before it and that confers on it some practical benefit. An *innovation* is something new, whether it be an invention, a way of doing business, or a collection of information, that *produces some economic value*. An invention, no matter how great a technical leap forward it represents, may have essentially no economic value, as you can see by looking at the scores of patents on inventions that have never produced a dime of revenue. Highly economically valuable innovations—such as new business models—may be based entirely on old technology. But the theory behind patent law is that granting patents on *inventions* is the best way to promote the use of those inventions in (economically valuable) innovations.

There are several reasons for this. One is the primary justification for patent law, namely that if the cost and risk of inventing a new application of a platform is high, and the ease and cost of copying it is low, then without patent protection (or some other form of reasonable prospective assurance of return on investment) innovators will not have sufficient incentives to innovate. We explored this argument in detail in Chapter 7.

There is another reason, however, that we have not yet explored. We typically think that competition is good at producing new technologies. Therefore we balk at closing off a newly discovered law of nature out of concern that such closure will allow only its discoverer to innovate using it, and thereby foreclose the kind of healthy competition in the field that we think will produce the most innovation. We could be wrong about this. After all, who are we to say that the discoverer of a new law of nature—the very person who had the kind of innovative spirit and rare ability to uncover a broad and deep fact about the universe that no one before had discovered—will not be just as successful at producing innovation in the field as all of his competitors combined if we just give him the patent rights he needs to shut them all out? Although we could be wrong, we don't think we are, and all of patent law rests on the assumption that we aren't.

We are inclined to allow patent rights on specific applications of laws of nature, natural phenomena, and other scientific facts, therefore, because we aren't as concerned that "application patents" will foreclose innovation as we are when it comes to "platform patents." If we grant a patent to someone on a new kind of automobile engine, and it turns out that the patent is too broad or should not have been granted at all, the harm done is minimized by the fact that others still have the freedom to design *other* car engines that don't overlap with the first car engine. The scope of an application patent is, by its very definition, limited to the particular application covered by the patent, and it leaves open the opportunity for others to develop different applications on the same platform. The ability to compete and thereby drive innovation is preserved, or at least not entirely foreclosed. Not so if we had granted a platform patent—say, on all transportation devices capable of traveling over roads, or on all devices based on the theory of relativity.

Our intuitions about what kinds of inventions patent law should and should not protect is therefore based not only on how we think innovation works in the real world but also on our own (well-placed) humility about our ability to draw the dividing line between patentable and unpatentable accurately in particular cases prospectively (ahead of time), and the risks of getting the answer wrong as the result of our poor skill at predicting the future. It is like objecting in general to the death penalty even though we agree that putting certain criminals to death is justified, for the reason that the harm caused by a wrong answer in even one case is too great to tolerate.

Open Innovation in the Age of Artificial Invention
In the Artificial Invention Age, artificial genies—such as genetic algorithms and neural networks—will be innovation platforms. Artificial invention technology is an innovation platform in the sense that it does not directly solve real-world problems but rather serves as a foundation for innovation by facilitating creation of products and services that do directly solve real-world problems. The Creativity Machine doesn't itself brush your teeth, but it enabled Stephen Thaler to invent a product that does.

It is easy to see that artificial inventions will be applications of artificial invention technology. Less apparent is the fact that artificial wishes—the fitness functions and other input that programmers input to artificial invention technology—will also be applications of artificial invention technology, in the sense that they are facilitated by and rely on artificial invention technology to achieve their creators' goals.

Therefore, if innovation is maximized when platforms are open and applications closed, then artificial genies should be open, and artificial wishes and inventions should be closed in the Artificial Invention Age.

This is generally consistent with the history of artificial invention technology to date. For example, the foundations of "evolutionary computation," of which genetic algorithms are one branch, have primarily been developed by academic and government researchers and then put into the public domain, where they are open.[17] Specific products and services produced using evolutionary computation, on the other hand, have been patented in many cases.[18] We can't, however, just sit back and expect this to continue; for the reasons I gave in Chapters 7 and 8 with respect to artificial inventions and wishes, we can expect private parties to seek patents on wishes and artificial invention technology more often and that such patents will be granted with increasing frequency. We've already seen patents being granted on advances in artificial invention technology itself. Therefore we will turn next to how patent law can be used to increase the likelihood, if not ensure, that the genie will be free but the wish bottled.

Using Patent Law to Free the Genie and Bottle the Wish
The Need for Flexibility
Before diving into specific recommendations for legal reform, it is worthwhile to note how daunting a task it will be to reform patent law to toe the dividing line between open platforms and closed applications in the Artificial Inven-

tion Age. So far I've made the problem seem relatively easy by relying primarily on our intuitions and some real-world examples. But as the old Yiddish saying goes, "'for example' is not proof."[19] Therefore we need to test our intuitions to make sure we don't adopt policies and broad rules on the basis of specific examples that are not generally applicable.

For example, you would be wrong if you thought we could free the genie and bottle the wish merely by prohibiting patents on genies and permitting them on wishes. If only it were so easy! The reason is that "open" does not mean the same as "unpatented" and "closed" the same as "patented." The same is true with any kind of property. The fact that you have a property right in a parcel of land does not necessarily make that land closed. You, as the property owner, might choose to make the land open for use by the general public without your permission, in which case the land would satisfy our definition of "open," even though it is proprietary.[20] You might even execute a legal document prohibiting yourself from going back on your word in the future. The Nature Conservancy, other environmental groups, and local governments often buy land, thereby making it proprietary, precisely to ensure that it is open to the public. "Proprietary" does not necessarily mean "closed."

The same is true with patent rights. A patent owner might choose to make its patented technology open to the public, for altruistic reasons, for public relations points, or simply to spur innovation in further technologies on the part of competitors (such as add-on products and services) that will produce even higher profits for the company itself than if it had kept the technology closed.[21]

Conversely, "open" is not coextensive with "unpatented" or "nonproprietary."[22] Some kinds of highly valuable business information may be unprotected by patent, copyright, or trade secret law for a variety of reasons, but the information might still be closed. Software that is distributed in an encrypted (scrambled) format is effectively closed under our definition if it is impossible or prohibitively costly for the public to examine, modify, and redistribute that software, even if no one owns any intellectual property rights in it.

The lack of correspondence between patented and closed on the one hand and between unpatented and open on the other hand means that no amount of tinkering with the rules governing what is and isn't patentable will automatically produce the right division between open platforms and closed applications under all circumstances. The same is true with changes we might make to other areas of law; private parties will—and should—always have leeway to work around the default rules imposed by the law. But we still have good reason to believe that patent law can do a reasonably good job at nudging the open-closed dividing line in the right direction, at least as a baseline. The law's lack of precision, however, is the first reason we need to ensure that the law remains *flexible* over time.

The second reason the law needs to remain flexible is that the dividing line between platform and application is both fuzzy and dynamic. Therefore any dividing line we fix into law will be inaccurate, and more so over time. Is a Web browser such as Microsoft Internet Explorer or Mozilla Firefox a platform or an application?

How about a word processor? Typically a word processor is called an "application" program because end users use it to write documents, not to create additional software. Yet most word processors include macro languages that allow their users to write programs called scripts to add new features to the word processor, such as printing postage or sending email. This makes the word processor look like more like a platform. Many "application" programs can also be platforms, insofar as they enable users to extend the features of the program or use it to create other programs. In fact, trace any application program backwards and you will find circuitry that provided a platform for a computer, which in turn constituted a platform for an operating system, all of which served as a platform for the application program. Each layer is both an application of the layer preceding it and a platform for the layer following it.

The third reason for maintaining flexibility in the law is the rapid rate of change in artificial invention technology itself and the difficulty of predicting the nature or extent of such change even a short distance into the future. Attempts to create laws tailored narrowly to specific technologies have often ended in failure. Complaints from the semiconductor industry in the United States that semiconductor designs were being knocked off by competitors led Congress to pass the Semiconductor Chip Protection Act of 1984,[23] which has seldom been used as developments in chip technology rapidly made the Act's provisions obsolete.[24] Patent law has done a better job by using (with few exceptions) general terms such as *machine* and *process* and leaving elaboration of the details to the courts and Patent Office. These lessons in the value of keeping the law flexible should give us pause before attempting to tie patent law too tightly to any particular dividing line between platform and application that is based on the current state of technology.

Finally, and perhaps most important, patent rights, like all property rights, at their best should create a playing field for efficient markets, not dictate their outcomes. Decreeing ahead of time that certain kinds of technology inherently constitute platforms and therefore cannot be patented, and that others inherently constitute applications that can be patented, will foreclose the use of those technologies as part of business models that might actually promote innovation better than others.[25] Just as a system of free speech should strive to ensure that no one is excluded from debates in the marketplace of ideas without deciding who the winner of any such debate is, patent law should strive to ensure that innovators are not precluded from using their inventions in business models of their own choosing as they engage in competition.

Therefore the challenge that lies ahead for patent law is to develop rules for promoting innovation in artificial invention technology by keeping platforms open and applications closed *by default*, while allowing the dividing line between platform and application to shift appropriately over time, and while affording sufficient flexibility for innovators to override the default rules as they see fit but without contradicting the basic policy of maximizing innovation.

Applying an Existing Framework

Given the difficulties of using patent law as a tool to free the genie and bottle the wish, we are fortunate that patent law already includes a longstanding mechanism that at least roughly fits the bill: the distinction between abstract ideas—which are not patentable—and practical applications of abstract ideas—which are patentable, so long as they are new, useful, and nonobvious and satisfy the other requirements for patent protection.[26] (Copyright law has a similar distinction between ideas, which are not copyrightable, and expressions of ideas, which are copyrightable.)

Although there are no precise definitions of "abstract ideas" and "practical applications," some clear cases should be illuminating. The equation $E = mc^2$ is an unpatentable abstract idea; a particle accelerator designed using this equation is a practical application of the idea, and therefore potentially patentable. Newton's second law of motion (which may be expressed using the equation $F = ma$) is an unpatentable abstract idea, while a jet engine operating in accordance with this law is a practical application of the idea.

Although this distinction has many justifications, one of them is consistent with our conclusion that keeping platforms open—to the extent that platforms are a kind of abstract idea—is the right way to promote innovation.

For example, in the *Morse* case we saw earlier, one of the reasons the U.S. Supreme Court considered claim 8 to be too broad was that:

> if this claim can be maintained, it matters not by what process or machinery the result is accomplished. For ought that we now know some future inventor, in the onward march of science, may discover a mode of writing or printing at a distance by means of the electric or galvanic current, without using any part of the process or combination set forth in the plaintiff's specification. His invention may be less complicated—less liable to get out of order—less expensive in construction, and in its operation. But yet if it is covered by this patent the inventor could not use it, nor the public have the benefit of it without the permission of this patentee.[27]

The Court, in other words, expressed the concern that if Morse's claim 8 were allowed and given a broad interpretation, it would stifle innovation by future inventors. The Court refused to grant Morse a platform patent on all possible applications of using electricity to transmit messages over arbitrary distances.

Similarly, a century later in the *Benson* case that we saw earlier, the Court struck down a software patent on a method for converting numbers from one format to another in part because mathematics is one of the "basic tools of scientific and technological work" and that to let the patent stand would effectively have been to allow a patent on a mathematical formula itself.[28] Although the reasoning in these two cases was somewhat muddled as applied to their facts, they are just two of many examples in which courts have expressed reluctance to grant broad patents out of an explicit concern that such patents would close off an entire platform for future innovation and thereby contradict the basic policy underlying patent law.

Erect High Hurdles

So how can we bring about a world in which the genie is free but the wish bottled? If artificial genies are the next big innovation platform, and if it so important to keep innovation platforms open, and if applying patent law's traditional rules to artificial invention technology will prove so difficult, why not just impose an outright ban on patents on artificial invention technology?

A flat ban on such patents, just like a ban on software patents generally, is not only too blunt an instrument but not even the most effective way to achieve open platforms. First, recall that the dividing line between platform and application is blurry and shifts over time. Although it may be true as a very general rule that artificial invention technology is an innovation platform, this may not be true in all places or at all times. Someone might apply for a relatively narrow patent on a small tweak to a genetic algorithm that is useful specifically to assist in the process of inventing new kinds of paper clips for holding between 2 and 20 sheets of paper. Is the new algorithm a platform or an application? There are good arguments for both positions; it is a platform for inventing paper clips within a narrow category, but also an application of genetic algorithm technology having a specific utility. This is yet another example where flexibility is the key, and where the actual outcome is best left to a patent examiner or court to decide on the basis of the particular facts of the case at that time. A flat ban on such patents could result in denying patent protection incorrectly in individual cases.

Second, although it may seem paradoxical at first, in some situations a technology, even a platform, is best kept open precisely by *allowing* it to be patented.

We can take this lesson directly from open source software. Some claim that open source software makes intellectual property obsolete, but this is exactly backwards. Intellectual property law—copyright law in particular—is the very *foundation* of open source software. Open source software could not exist without copyright law.[29]

How can this be? Doesn't copyright law give copyright owners the power to *prohibit* others from copying their works, while the purpose of open source is to ensure that anyone can copy open source software *without* obtaining permission from anyone else?

To understand why open source needs copyright law, imagine the following scenario. You create a new operating system and want to distribute it as open source software, which means you want to ensure that anyone who receives a copy of the operating system can redistribute it, including any modifications they make to it, without receiving permission from you or from anyone else. In an attempt to do this, you distribute the operating system along with a document (which looks just like a common open source license) stating generally that anyone who receives a copy of the operating system does not need permission from you or anyone else to redistribute it or modified versions of it.[30]

Someone else—call him Mr. Closed—has other ideas. He wants to free-ride on your labor and sell a proprietary version of your operating system. So he takes your operating system and adds some features to it. He obtains a copyright on his modified version of the operating system. Now he can pro-hibit anyone—including you!—from using or making copies of the modified operating system without his permission. This is called "going closed," and it is what every open source software developer strives to prevent from happen-ing to her software.

What action can you take against Mr. Closed? The answer is absolutely nothing, *if* you did not have a copyright in your original operating system. If you do not have a copyright, then you have no legal basis to stop Mr. Closed from doing anything he wants to your dear operating system, including making new versions of it proprietary against your wishes. If, however, you have a valid copyright in the original operating system, then you can sue Mr. Closed for copyright infringement because he is now distributing copies of portions of your copyrighted work without your permission—the heart of a copyright claim.

So not only are open source and copyright not in contradiction with each other, and not only does open source not make copyright obsolete, but copy-right is the *foundation* of open source. The latter would have no teeth without the former. Richard Stallman, the founder of the free software movement, recognized this when he coined "copyleft," referring to the use of copyright to ensure that modified versions of open source software are kept open. He used the term to reflect this kind of "legal jujitsu" in which copyright protection is obtained but then turned around and used to ensure openness rather than the private control copyright law directly provides.[31]

The same is true of open innovation in the context of artificial invention. If we want to give innovators the ability to make their newer and better genies open in the same sense as open source software, which means giving them the ability to stop others from making improvements on their innovations closed (proprietary), then we need to allow such innovators to obtain intellectual property rights—namely, patent rights—in those genies if they are truly to be open platforms. Therefore, although at first glance a ban on artificial inven-tion patents would promote open platforms, such a ban would in fact make it more difficult for innovators to ensure that the artificial invention platforms they create remain open.

This does not imply that *all* genies should be patentable. As with all in-tellectual property rights, the key is *balance*. Although allowing patents on genies may be necessary in some situations to ensure that such genies can be maintained as open platforms, allowing such patents also creates the opportunity for individual inventors to grab patents on key artificial invention technologies and use them to lock others out of the market unreasonably, or in ways that otherwise stifle innovation. The best defense against this is to ensure that patent law's rules are interpreted strictly, at least to guarantee that the patents granted are justified and not interpreted too broadly.

Strictly Interpret Existing Standards

The best way to be sure patent law continues to maintain the dividing line between unpatentable abstract ideas and patentable practical applications of them is not to attempt to redefine or directly interpret those terms themselves, but rather to update and strictly enforce patent law's other existing rules in light of artificial invention technology. Particular examples will demonstrate this best.

I already explained in Chapters 7 and 8 how we can update patent law's nonobviousness and disclosure requirements in the context of patents on artificial inventions and wishes, respectively. The same attention to detail should be applied to patents on genies. An improvement to artificial invention technology, such as a modified kind of genetic algorithm, should be patentable only if the inventor describes in sufficient detail how to make and use it employing ordinary skill in the art (that is, without the need to engage in what patent law calls "undue experimentation") and the improvement would not have been obvious to a person having ordinary skill in the art of artificial invention. Although the question of whether artificial genies should be patentable looked at first like a more difficult problem than the rest, so far we need only apply existing rules of patent law to them. In this sense, applying patent law to genies turns out to be much easier than applying it to artificial inventions or wishes.

These rules will help to maintain the correct dividing line between unpatentable abstract ideas and patentable practical applications. If an inventor's description of a new kind of genetic algorithm does not adequately describe how to make and use it, this might be evidence that the inventor has not actually invented any new algorithm, or that the inventor's work product so far is too abstract to produce actual working software. In this case, the patent application should be rejected for failing to satisfy the enablement requirement, but as a side-effect this will prevent the patent from being granted on an abstract idea. The more abstract the description, the more likely it is that the patent application will be rejected, and rightfully so.

Even more important, however, will be patent law's utility requirement, which we have largely ignored so far. To satisfy the utility requirement an invention must be useful. This requirement received little attention until recent years. The leading treatise on patents, by Donald Chisum, states that "an invention must be capable of some beneficial use."[32] As you can imagine, it is not difficult for a smart inventor or patent lawyer to write a patent application for essentially any invention that satisfies this definition of utility.

The utility requirement has been beefed up in recent years in response to biotechnology and software patents that have pushed its boundaries. For example, the Patent Office has issued guidelines stating that one could not satisfy the utility requirement merely by providing an incredulous use for one's invention, "such as the use of a complex invention as landfill."[33] If such utilities were considered sufficient, then *any* physical object would satisfy the utility requirement. Now inventors must prove that their inventions have a *substantial*, *specific*, and *credible* utility.[34] The precise meaning of this definition isn't important for our purposes; what is important is that it represented an effort by the Patent Office to require more proof from inventors that their inventions could actually do something directly useful in the real world.

This only begs the question: what kinds of uses—practical applications—of software should qualify as substantial, specific, and credible? The easy cases are those in which software performs the same function as some old-fashioned and clearly useful device, such as controlling the brakes in a car. But the more abstract the purpose of the software, the more difficult it is to tell whether the software is sufficiently useful. For example, if the purpose of a particular computer program is to calculate the results of a new mathematical formula, should the program satisfy the utility requirement? What if the patent application for the software points out a specific, narrow, practical application of the software and therefore of the formula, such as determining how frequently to pump the brakes on a car?

The reason applying the utility requirement to software has proven so difficult, and the reason it will continue to prove difficult as applied to artificial genies, is that before the advent of computers courts did not have to face squarely the question of whether a machine performing a new and nonobvious mathematical calculation could qualify for patent protection if the act of embodying the calculation in a machine involved only ordinary skill. Prior to the advent of computers, if you developed a new formula of any substantial complexity and wanted a machine to carry out the formula, you would need to custom-design some machinery to do so. The legal system, therefore, focused on whether the *machinery* was novel and nonobvious, and no one questioned that such custom-designed machinery was useful even though all it did was calculate a formula.

Now, if you develop a new formula and program a computer to calculate it, the resulting programmed computer performs exactly the same function as the custom-designed formula-calculating machine of yesterday. Yet now we question whether the formula, as programmed into the computer, satisfies the utility requirement. Perhaps we should have questioned this all along, but we only thought to do so once it became easy to create a formula-calculating machine merely by programming a computer.

Yet we must answer this question if we are to decide whether a mathematical formula, as programmed on a computer, can be patented, because neither of the other primary requirements for patentability—novelty and nonobviousness—can necessarily rule it out. Mathematical formulas certainly can be new, and mathematicians would certainly find some of them nonobvious. The programmed computer is certainly a machine, which is one of the categories under which an invention must fall according to U.S. patent law if it is to be patentable. Therefore if formulas expressed in computer programs are to be excluded from patentability, this must be due to their being not useful. The same is true of *any* information that can be stored in a computer, such as digital music or art.

You might think that a mere mathematical formula must fail the utility test because it does not have the kind of practical utility that patent law requires, even if the formula is programmed into a computer; but the conclusion is not so easy to reach as you might think. If the new formula is anything other than a pure theoretical curiosity and has any remotely valuable features, even in pure mathematics, an inventor working with a savvy patent lawyer can almost always devise some plausible practical use for the formula if it is calculated by a computer rather than mentally or by hand. The use might be as simple as enabling the formula to be computed more efficiently, even if no one has discovered whether those computations can be applied to real-world problems. But this only begs the question: Is enabling calculations to be performed more efficiently by a computer a practical use that should be protectable by patent law? If not, why not?

In many cases, it is possible to go further and make a much stronger case for patentability. If the formula in question has any application in physics or engineering, such as clarifying tumors in an x-ray image, the case for practical utility becomes even stronger.

Getting back to the topic of this chapter, does a genie have practical utility? On the one hand, a genie has the most powerful kind of practical utility possible: the ability to produce any invention that can be generated using a wish you can describe in the magical language of wishes. On the other hand, a genie's power is too abstract to be patentable, in the sense that it does not solve real-world problems *directly*, but only indirectly by enabling inventors to create inventions that solve such problems.

Patent law traditionally protects inventions having this kind of indirect utility, at least once they have been demonstrated to be actually capable of producing something of value. This was the patent applicant's problem in *In re Fisher*, a recent biotechnology patent case in which a patent application that listed *seven* distinct utilities for the invention was nonetheless rejected on the grounds that all of the uses alleged by the inventor "represent merely hypothetical possibilities, objectives which the [invention] could possibly achieve, but none for which they have been used in the real world."[35] Unlike a microscope or other patentable research tool, Fisher had not demonstrated that his invention could be used, even if only indirectly, to produce practical, real-world results.

Patent law should treat artificial genies in the same way. A patent on a new kind of artificial genie should be granted only if the inventor has described in sufficient detail how to make and use that genie, and if the inventor has demonstrated that the new genie can be used to produce a new invention, or produce existing inventions more efficiently. The burden of proof should be on the inventor.

Once the inventor has satisfied this burden of proof, the genie should no longer be considered an abstract idea. New kinds of genies that can demonstrably be implemented in computer software or in some other kind of machine, and that have some proven ability to produce new inventions or enable inventing to be performed more efficiently or effectively, are not abstract ideas and have at least as much practical utility as any other kind of device that facilitates inventing, from a hammer to a microscope to a computer. The new challenge that will arise, as we shall see in Chapter 10, is how to administer these requirements in practice.

Another problem, which I can only point out but not solve, is this. Consider a genie that is a genetic algorithm shown to evolve an artificial neural network that in turn has produced a program capable of generating a design for a new paperclip. In such a case, utility has been established, no matter how indirect the link between the genie and the resulting real-world invention. But now imagine your genie, a genetic algorithm, has evolved a novel and nonobvious artificial neural network that has not been used to produce anything in the real world. Is this genie sufficiently useful, even though it has not produced any physical products? Is advancing the state of knowledge of artificial neural networks a sufficiently practical utility? This is one question the law must struggle with case by case.

Note also that not all new digital genies will necessarily be patentable, even if they have practical utility. They must still be nonobvious. As I argued in Chapter 7, advances in artificial invention technology itself are sure to raise the bar for nonobviousness, thereby acting as another safeguard against granting artificial invention patents on abstract ideas.

Allowing patents of this kind on artificial genies will also not necessarily grant protection to entire platforms. A claim to a particular new kind of genie will cover only that genie, not others to be invented in the future. Therefore any particular genie patent will still leave room for other inventors to create their own genies.

Furthermore, a patent on a genie will not necessarily cover wishes written for the genie, any more than a patent on computer hardware grants its owner protection for software written on the computer, or a patent on a piano grants copyright protection in songs written using the instrument. The result I recommend therefore preserves intellectual property law's traditional balance between inventors of devices and inventors who create additional inventions using those devices.

The Law Is Not Enough

We do not need to ban patents on artificial invention technology or create new legal rules in light of it. Patent law's existing rules possess sufficient flexibility to adapt to artificial genies. We need, however, to reinterpret such rules in light of artificial invention technology to ensure that they continue to achieve their original goals, just as a butcher recalibrates a scale after it has been in long use to ensure that it continues to measure weights accurately.

In particular, we need to strictly apply patent law's utility requirement and its distinction between abstract ideas and practical applications. These legal rules will therefore protect against the worst kind of abuses that could result in overly broad or otherwise unjustified patents being granted on new platforms for innovation.

Patent law will still allow patents for artificial genies, and it is possible that some of these patents will in effect be patents on platforms. If platforms should be open, then won't this allow a small number of patent owners to keep platforms closed and thereby undercut the basis for innovation I've described?

Not if everything we've seen about open innovation is correct. The Ethernet protocol was patented and yet became an open standard. As more innovators realize the benefits of open innovation, not just for innovation's sake but for their own bottom lines, even patented genies are not likely to be used as mere sledgehammers to beat down the competition.

More generally, we cannot expect patent law itself to ensure that all platforms remain open. Rather, it will take a combination of patent rights, market forces, and collective action to provide a web of protections for innovation. As an example of what I mean by "collective action," companies have recently begun to join forces to contribute their software patents to so-called open source patent pools.[36] The idea is that if you contribute your software patents to a particular pool, you are immune from being sued on *other* patents in the pool. Patent pools are a way for companies to use their patents *defensively* and to assure others they won't use them *offensively*. Open source patent pools are a relatively recent development and are still in the experimental stage. Whether or not they work, however, they demonstrate that open versus proprietary is a false dichotomy; open source software is *both* open *and* proprietary, and sometimes software can best be kept open by leveraging property rights in it.

As I stated at the beginning of this chapter, patent law should strive to create a playing field in which innovators can experiment with diverse business models, whether they be open, closed, or a combination of the two. At the very least, we can strive to ensure that patent law does not inadvertently close off any of these options to innovators.

As powerful as patent law is, it cannot do everything. Nor can it even hope to achieve its limited goals for long and without great effort and commitment, as we shall see next.

10 The Limits of Law in a World of Genies

EVEN IF ALL OF THE PROPOSALS for legal reform that I made in the previous chapters were enacted tomorrow, patent law and artificial invention technology would coexist peacefully for only a limited period of time. Laws, like all human creations, are developed based on assumptions about the context in which they will be applied; as such, laws serve their intended goals only to the extent that those underlying assumptions continue to hold true. A car designed to travel at speeds up to 80 mph will begin to rattle at 100 mph and fall apart at 150 mph; even revamping the engine from the ground up might get you to 250 mph but not 10,000 mph. Similarly, patent law, which was developed to provide incentives to invent in a world in which inventing was risky, time-consuming, and expensive, will bear only so much reduction in inventive risk, time, and cost before requiring more radical restructuring.

Those who make and interpret laws have long been aware of this phenomenon. Farsighted lawmakers attempt to craft laws to be flexible in response to shifting circumstances. This is why we call the U.S. Constitution a "living document"; its meaning changes to reflect the changing world around it. Even the most flexible legal system, however, can withstand only so much change. We now explore the challenges that the patent system will confront in the face of increasingly powerful artificial invention technology, and the steps that can be taken to extend patent law's useful life as far as possible into the future.

Bridging the Gap Between Wish and Invention

Recall that patents are documents intended to be understandable to the public (although you may find this hard to believe if you've ever tried reading a patent). A patent is intended to inform the public about (1) how to *make* and *use* the invention covered by the patent, just as an architect provides a builder with a blueprint for constructing a building[1]; and (2) the *legal rights* that the patent confers on the patent's owner, just as a deed to a parcel of real property points out the "metes and bounds" of the property. If you are an automotive engineer of ordinary skill, you should be able to learn how to make and use a particular patented car frame just by reading the main text of the patent, called the "specification."[2] Similarly, you (or at least your patent lawyer) should be able to learn which car frames the patent prohibits you from making by reading the claims of the patent.

This understanding of how patents are supposed to work rests on assumptions about the ability of a *human* to absorb and draw conclusions from the information in a patent. We don't learn about patented inventions by examining those inventions directly. Instead, as shown in Figure 11, we learn about patented inventions by reading the corresponding patents and drawing inferences from them.

The patent document is a proxy for the "invention itself" and an *intermediary* between the invention and our minds. The patent system doesn't have to work this way; we could eliminate patent documents entirely and just require inventors to submit working models of their inventions to the Patent Office

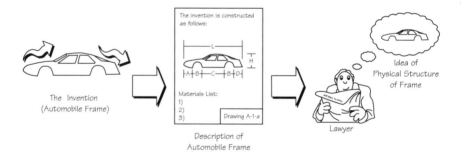

FIGURE 11 From invention to description to understanding.
Courtesy Spinney Associates

as proof of what they have invented.[3] Any member of the public who wanted to know what has been patented could then examine those models and draw conclusions from them directly, without inserting a *description* of each model as an intermediary between the model and human observer.

The existing document-based patent system works well enough for patents on car frames and other inventions invented the old-fashioned way, which result in patent documents directly describing the physical structure of the invention and how to make and use it. A human mind can work backwards from the text of such a patent to the invention itself, thereby bridging the gap between the description and the invention.

Artificial invention technology is widening this gap between description and invention by introducing artificial genies between the two. This has two significant consequences:

1   In some cases we may find it difficult or impossible to understand inventions produced using artificial genies by studying those inventions directly with our unaided mind; we can only understand them in terms of the wishes that produced them. Traditional software is a perfect example. Once you write source code for a computer program of any substantial size and compile it into machine code, even an expert programmer won't be able to understand the software by directly examining the electrical signals of which it consists. Instead, she must go back to the source code—the wish—as her guide to understanding the software.

2   Conversely, when evolutionary computation and other nondeterministic forms of invention automation technology are used, merely *reading* a wish may not enable us to predict which invention(s) the wish will produce when an artificial genie grants it. In such cases, the only reliable way to discover which inventions the wish will produce is to run it through an artificial genie and observe the results.

In both cases, patent law's reliance on the human mind to understand the invention on the basis of a description of the invention breaks down. One practical consequence of this breakdown is that when the Patent Office receives a patent application for a wish, the Office will not easily be able to ascertain the *field of technology* ("art") to which the patent application relates. The Patent Office traditionally identifies the field of a patent application by having a human employee skim briefly through the text of the patent application; if, say, it claims an "automobile frame," the employee determines that automotive engineering is the field of the application. The Office then assigns the application to a patent examiner who has a background in automotive engineering and who specializes in examining automobile frame patents. This process produces accurate results if the field of the invention can be gleaned easily from the patent's description of the invention.

If, however, the patent merely claims "any device produced using the following airflow equation," the Patent Office may have no way of knowing—merely by reading such a wish—whether presenting the airflow equation to an artificial genie will produce an automobile frame, an airplane wing, a football, or all of the above. As a result, it may be difficult or impossible to select a patent examiner with the right technical background to examine the patent application.

The Patent Office identifies the field of each patent application not only to pick the right examiner for the job but also to let the examiner know where to look for "prior art"—previous inventions that are similar to the one covered by the application on her desk. How would a patent examiner use our airflow wish patent application to figure out which types of prior art to search for? Should he search for existing airflow equations? car frames? airplane wings? footballs? If we know he should search for car frames, how is he to draw this conclusion just from reading about the airflow wish itself, particularly if the patent application says nothing about the car frames the wish can produce?

These problems result directly from the introduction of powerful genies between descriptions of inventions and inventions themselves, because such genies can cause the field of the description to differ significantly from the field of the resulting invention. An inventor may have used techniques in the art of mathematics to write a wish that produced an invention in the art of automotive engineering. As a result, a description of the former may not give any clue as to the field of the latter, and vice versa.

Similar problems arise for patents on artificial inventions themselves. Consider a case in which an inventor uses a wish to produce an artificial invention—say, the NASA antenna we saw in the Introduction, which the inventor himself does not understand. He knows *that* it works, but not *why* it works. He *does* understand, however, the wish that he used to produce the antenna. In the patent application, he describes in great detail how to build the antenna and supplies experimental data proving that it works. He does not, however, describe the wish. He has done nothing impermissible under traditional principles of patent law, which only require him to describe *how* to make and use his invention, not to explain *why* it works.

Once again, the Patent Office will find it difficult to identify the art of this patent application. The examiner may read the patent application and—because it talks only about antennas—think simply to search for other antennas having similar shapes. She might not think to look for wishes that could have been used to produce the antenna. Doing so would require the examiner to work *backwards* from the specific antenna described in the patent application to the categories of wishes that might possibly have been used to produce it. This could be an impossible task, given how difficult it can be even to work *forward* from a known wish to the antenna(s) it could produce.

Although the Patent Office might respond to this problem by running the airflow equation through an artificial genie itself and seeing what comes out the other end, doing so would impose a significant burden on the Office. The Office could consider such patent applications to the fall within the field (art) of "artificial invention technology," but this solution would break down as the number and percentage of applications assigned to this field grows.

Members of the public may find it difficult to determine which specific products are covered by a wish patent's claims, for the same reasons. Requiring the public to run a patented wish through an artificial genie to identify which products are covered by the patent may not be overly burdensome if the right artificial genie is readily available to the public. (If it wasn't, the wish patent shouldn't have been granted in the first place.)[4] What makes the problem trickier is determining which *version* of artificial genie you must apply to the wish. Remember that applying the same wish to tomorrow's artificial genie may produce a wider range of inventions than applying the wish to today's artificial genie. As artificial genies grow more powerful, they tend to produce better results using the same wishes. Even if we decide that the right genie is the one in existence at the time that, say, a particular wish patent was granted, then determining which specific inventions the wish patent covers would require you to obtain an artificial genie extant on the day the wish patent was granted and *then* run the wish through it to see what comes out the other end, just to know which specific inventions the wish patent does and does not cover.

By freeing us from the need to engage directly in the gruntwork of producing specific inventions or even understanding how those inventions work, artificial genies may leave us increasingly unable to understand those inventions directly, and therefore make us more *reliant on genies* for such understanding, just as Henry Ford instructed his team of engineers to "design an engine with all eight cylinders cast in one block" and then had to rely on those engineers to explain to him how the V8 engine they designed worked.

Although patent law has dealt before with inventions that cannot be understood directly, patent law's previous solutions will not work for all artificial invention technology. Chemicals have long been patentable even though a chemist may not be able to describe the chemical structure of a chemical C he has invented by combining chemicals A and B. The best he may be able to do is to tell you that chemical C is "the substance you obtain when you combine chemicals A and B." Patent law allows chemical inventors to both describe and claim the chemicals they invent in terms of the process of making them, in what are called product-by-process patents. The description in such a patent of the process for making a chemical can be sufficient both to teach the public how to make and use the invention and to point out clearly which chemicals the patent does and does not cover.

The reason we can't simply apply this solution to artificial invention technology is that wishes are multiply realizable, as we learned in Chapter 5. Recall that this means you can give a single wish to an artificial genie to produce a variety of specific inventions satisfying the wish. Not so in our chemical example: follow the chemist's instructions anytime, anywhere, and you will always obtain the same chemical. The chemical you obtain doesn't depend on our understanding of chemistry or change as chemical technology improves. Therefore, although allowing chemical inventors to describe and claim the chemicals they invent in terms of the *process used to make them* does impose some additional burden on the public by requiring us to follow the chemist's instructions to obtain and then observe the results if we desire to study the chemical itself, we only have to do this *once*, and after having done so we know forever exactly what the patent covers. None of this is true for wish patents because of multiple realizability. As a result, the cost imposed on the public by patents on wishes and artificial inventions could be much more substantial, making patent law's traditional product-by-process type of solution inadequate to the task.

The multiple realizability of wishes creates another circumstance that differs from those with which patent law has had to grapple in the past. In most cases, broad and sweeping claims to an entire class of devices defined in terms of the function they perform, such as Morse's claim on all electrical message-transmitting devices, could be rejected on the grounds that the inventor had not actually tendered a description that a person having ordinary skill in the art could apply to make and use the entire class of devices. In contrast, inventors of wishes will be able to back up their claims to broad wishes because increasingly powerful artificial genies will be able to handle the gap-filling legwork that Morse lacked. A wish patent, therefore, may yield a broad description that enables the public to make and use a wide range of devices, but not serve the public notice function because merely reading the wish claim does not convey to a human reader what the claim covers.

Whenever changed circumstances create costs that cannot be avoided, we must decide how to *allocate* those costs. The enablement and public notice functions of patents suggest that we shift these costs primarily to patent applicants, such as by requiring inventors of wishes and artificial inventions to submit more detailed descriptions and claims in their patents. Biotechnology patent applicants already are required to describe their inventions in great detail because of the unpredictability of biotechnology.[5] For the same reasons, it may be desirable to require wish inventors to describe experimental results demonstrating that the wishes for which they are seeking patent protection actually have been used to produce a range of specific inventions falling within the scope of the wish.[6] In the case of patents on artificial inventions, inventors should be required to disclose the wishes and artificial genies they used to produce those inventions, if the inventions themselves are not readily understandable on their own terms.[7]

One pitfall to avoid, however, is jumping too quickly to the conclusion that more detail necessarily is better. Recall from Chapter 8 that a voluminous patent describing 1,000 car frames may actually be *more* difficult for the public to understand than a short, abstract patent describing the wish used to produce those frames. Recommendations that we require programmers to submit detailed source code as part of their software patent applications are flawed for the same reason; in many cases doing so would result in convoluted patents requiring *more* work to understand than a concise, well-written patent application without source code. The key will be to remain flexible and attentive to the solutions that best implement the goals of the patent system at a particular time according to the state of technology at that time.

Many of the problems we've just explored, which will be exacerbated by artificial invention technology as it develops, stem not from flaws in the fundamental principles of patent law but from particular ways in which those principles currently are applied in practice. The right way to fix this kind of problem is to reinterpret existing law so it reflects how artificial invention technology affects the process of invention. Once again, our experience with software patents points the way to many such reinterpretations that may be useful as starting points for consideration in the Artificial Invention Age, such as affording patent examiners increased access to outside technical experts, shortening the term of software patents (to perhaps two to seven years),[8] creating a limited right to reverse-engineer patented software,[9] importing a copyrightlike "independent invention" defense into patent law,[10] allowing patent protection only for those software inventions that actually are embodied in commercial products,[11] eliminating or modifying the patent examination process to enable software patents to be issued more quickly,[12] and establishing a compulsory licensing scheme for software patents.[13] These proposed reforms give just a flavor of the ways in which aspects of a patent system can be tweaked to address particular problems caused by software,[14] and that could also be helpful in the context of artificial inventions. The U.S. Patent Office has already adopted some measures in response to criticisms of business method patents addressing at least some of the concerns about software patents and artificial invention patents expressed here.[15] One should not underestimate the extent to which such changes, although appearing individually to be relatively minor, can help to restore the delicate balance patent law is intended to maintain.

The Vanishing Pillars of Patent Law

Patent law will eventually break down, even if we reform it, because artificial invention technology is undermining the bases for three distinctions that form the foundation of patent law, namely the distinctions between (1) different arts (fields of technology), (2) basic science and applied science, and (3) the liberal arts and the sciences (or "useful arts").

The distinction among arts is reflected most prominently in division of the Patent Office into "art units": groups of patent examiners, each of which is assigned to examine patents in a particular art. One art unit examines only patents for baby diapers, another only for toasters, and never the twain shall meet (let's hope). Similarly, the Patent Office maintains a system for classifying technologies.[16] Each art unit examines patents only within a certain class or small number of classes. This helps the Patent Office examine patent applications more efficiently and effectively, because it can hire examiners with the right technical background to work in each art unit. If it were not possible to classify patent applications arriving at the Patent Office's door within a particular art, the quality of patent examination would suffer greatly; there would be no choice but to assign patent applications to patent examiners lacking the right technical background to examine them.

Next consider the distinction between basic science and applied science, sometimes referred to as the difference between science and technology. We've already seen the important role this distinction plays in patent law; it constitutes the basis for the distinction among the three categories of *per se* unpatentable subject matter (abstract ideas, laws of nature, and natural phenomena) and patentable practical applications of those ideas, laws, and phenomena. Although scientific principles are not patentable, real-world machines embodying those principles are patentable. If this distinction were to disappear or become impossible to enforce in practice, we would have no choice but to allow *anything* that is new, useful, and nonobvious to be patented, or else eliminate patents altogether.

Finally, consider the distinction between what we now call the liberal arts and the sciences or useful arts. It forms the basis of the distinction between human-made works that are protected by copyright law and those that are protected by patent law. Works in the liberal arts, such as novels, poems, songs, and plays, are susceptible to protection by copyright law. Works in the useful arts, such as machines, articles of manufacture, compositions of matter, and industrial processes are susceptible to protection by patent law. Although copyright and patent law are fundamentally similar in that both confer exclusive rights for a limited period of time to encourage rights holders to create new works and disclose those works to the public, they differ significantly in how they *implement* those rights. Copyright law includes certain defenses (or exceptions to infringement), such as fair use and independent creation, out of concern for freedom of speech.[17] Patent law includes no such defenses. Copyrights last for several decades longer than patent rights.[18] The list goes on. Therefore, which of these two types of legal rights attaches to any particular work can have a significant impact on how available the work is for use by the public. If we could not tell whether a particular type of work fell into the liberal arts or the useful arts, then we would have no basis for deciding whether to apply copyright law, patent law, some combination of both, or neither. Computer programs—the first kind of artificial wish—engendered exactly such a debate about the appropriate form of legal protection, a debate that continues to this day with no clear end in sight.[19]

Artificial invention technology is eroding these three distinctions. We've already seen how such technology is blurring the boundaries among different "arts" by making it increasingly possible to use techniques from one art, such as physics or computer science, to write a wish that an artificial genie can automatically transform into a design, such as an antenna or automobile frame, falling within another art. To the extent that the Patent Office and courts will have difficulty not only in identifying the right art to apply to a particular invention but also in understanding and applying the teachings of those arts, it may be useful for those institutions to draw more frequently and systematically on members of the technical community for assistance. Computer Science Professor Jeffrey Ullman, recognizing the seeds of this problem in relation to software patents, has urged computer scientists to become more active in lending their expertise to the patent system to help evaluate whether particular software patents should be granted.[20]

More recently, the U.S. Patent Office instituted a pilot project, called Peer to Patent, inspired by the peer review process used to evaluate scientific papers.[21] The basic motivation behind the program, originally proposed and spearheaded by Beth Noveck of New York Law School, is to enable patent examiners to receive input from technical experts (and the public more generally) on individual patent applications to assist the examiners in determining whether the inventions they are examining are novel and nonobvious. The merits of this or any other particular project are not important. Rather, the point is that projects enabling the Patent Office to obtain input from technical experts on a just-in-time basis may be a more efficient and effective way for the Office to incorporate the latest knowledge from a variety of arts into the examinations it performs, particularly if those arts and the borders between them change rapidly.

The blurring distinction among different arts will also affect how courts decide disputes over patents in the Artificial Invention Age. In the United States, for example, the Federal Circuit Court of Appeals, which decides all patent appeals, has developed standards for biotechnology patents distinct from those for software patents, on the basis of assumptions about how inventions in each field are created.[22] For example, the Federal Circuit has required biotech inventors to describe their inventions in great detail, on the assumption that without such detail it would not be possible to manufacture such inventions. This may not be true in the Artificial Invention Age, however, when a high-level description of a drug may be sufficient to produce it. The Federal Circuit has also tended to find biotech inventions to be nonobvious, based on the assumption that biotechnology is unpredictable and that finding new biological structures is difficult.[23] Such an assumption may not be valid in the Artificial Invention Age, as search-based automation techniques become ever more capable of finding predictable solutions to increasingly complex problems.[24]

Conversely, the Federal Circuit has allowed software inventors to obtain software patents incorporating relatively abstract descriptions of their inventions, on the premise that such descriptions are sufficient to enable other programmers to recreate the invention with little experimentation. This may not be true for wish patents; merely submitting a description of a wish may not be sufficient to recreate the problem solution without also specifying the genie that was used to grant the wish and the initial conditions under which the wish was granted. The Federal Circuit has also indicated it is inclined to find software patents obvious in light of the relative ease of innovating in software. Such an assumption may no longer be valid in the Artificial Invention Age as programmers use highly interactive and experiment-based development techniques to design complex software.[25]

Such premises may have had some basis when fields such as biotechnology and software employed distinctly different modes of inventing and manufacturing, each associated with its own economic patterns of innovation. For example, it is still safe to say that inventing a drug is orders of magnitude more expensive than inventing a new piece of software (even if the software is Windows Vista). But such distinctions are already showing signs of breaking down.[26] The cost of the Human Genome Project by its completion in 2003, for example, was nearly $3 billion. Just a few years later, using the same technology, decoding a human genome cost only about $25–50 million. By the end of 2008 the cost was expected to be as low as $100,000.[27] At this rate it may not be long before we see "garage biotech" startups that mirror the computer hardware and software startups that sprang up beginning in the 1960s. "Retail genomics" companies are already appearing, which will sample your DNA and inform you of the diseases for which you are most at risk, for as little as $1,000.[28] Supercomputers such as the Japanese Protein Explorer are being used to test thousands of chemical compounds for their ability to bind to proteins in the human body and therefore potentially serve as new drugs, avoiding traditional time-consuming, expensive, and messy laboratory experimentation.[29]

It will make no sense to maintain differing legal rules for numerous "arts" in the face of such developments, as the techniques and technologies that are used in inventing bleed from one field to another, and as economic patterns of innovation no longer break down cleanly along the lines of traditional technology industries.

The second distinction artificial invention technology will blur is the one between science and technology. Let me make clear what I mean when I say that artificial inventions will "blur" this distinction. A law of nature will never *be* a machine; gravity will never *be* an airplane, even in the Artificial Invention Age. What an artificial genie can do is reduce the time, cost, and effort required to *transform a description* of a scientific principle into a machine that *applies* the principle. We mean the same thing when we say that cell phones blur the line between home and work; although your house is still physically distinct from your office, the ease with which co-workers can contact you by phone makes it easier to perform worklike functions while physically at home. Similarly, artificial genies will make it easier for someone who discovers a new scientific principle to produce a machine that operates according to the principle. In this way, artificial genies will blur the line between science and technology.

A similar blurring of the distinction between instructions for building a device and the physical device itself is already starting to affect the patent system. AT&T owned a patent[30] on techniques for enhancing the sound quality of computer-generated speech. Microsoft incorporated software into its Windows operating system that infringed AT&T's patent.[31] Microsoft created a small number of "golden master" disks of its Windows operating system in the United States and then sent those disks overseas to foreign computer manufacturers, who installed the software from that small number of golden masters onto thousands of computers sold outside of the United States.[32] AT&T sued Microsoft for patent infringement. The dispute focused on whether Microsoft's act of sending the golden masters overseas constituted only several acts of infringement in the United States or *thousands* of acts of infringement under a particular section of the U.S. patent statute that prohibits "suppl[ying] . . . from the United States all . . . of the components of a patented invention."[33] Both parties cared deeply about the answer to this question because it would determine whether Microsoft had to pay AT&T for a handful of infringing acts or for thousands of them.

Microsoft argued, in essence, that the master disks were not "components" of the thousands of overseas computers within the meaning of the patent statute because those disks were not—and could not—be physically installed in thousands of computers overseas. A few disks cannot be installed in thousands of computers in the same way that thousands of graphics cards can be installed in thousands of computers, one card per computer. Microsoft argued that the disks themselves were not "components" of the overseas computers because it was not the *disks* that were installed on the overseas computers, but rather *software copied from those disks.* Microsoft further argued that the true "components" of the overseas computers were the *copies* of the software, that such copies were made overseas, and that Microsoft therefore had not "supplied" those copies from the United States as required by the patent statute.[34] The Federal Circuit disagreed and held Microsoft liable for every overseas copy:

> Given the nature of the technology, the "supplying" of software commonly involves generating a copy. For example, when a user downloads software from a server on the Internet, the server "supplies" the software to the user's computer by transmitting an exact copy. Uploading a single copy to the server is sufficient to allow any number of exact copies to be downloaded, and hence "supplied." Copying, therefore, is part and parcel of software distribution. Accordingly, for software "components," the act of copying is subsumed in the act of "supplying," such that sending a single copy abroad with the intent that it be replicated invokes [patent infringement] liability for those foreign-made copies.[35]

Whether the Court of Appeals reached the right conclusion is not important for our purposes. What is important is that, before the advent of programmable computers, there was a relatively bright line between *instructions for building a patented component and the component itself.* A blueprint for a house clearly is not a house. At first glance, the disks that Microsoft shipped outside the United States also appear quite different from the computer memory into which the software was copied. Yet you can appreciate the intuitive appeal of concluding that Microsoft performed the *equivalent* of shipping thousands of copies of the software overseas even though in fact the company shipped only *a few* copies, because the software on each disk was like a set of instructions for copying itself, and because the overseas computers were capable of *automatically following those instructions to copy the software at essentially no cost.* This is what I mean when I say that artificial genies, which can automatically follow instructions to create software and other products, are blurring the line between abstract instructions and concrete products.

This line-drawing problem will only become more widespread as a broader range of products are "distributed" not by shipping the completed products on planes and trains but rather by transmitting *computer-readable instructions for manufacturing those products*, and then manufacturing the products themselves on-site and on-demand. Jordan Pollack of Brandeis University described to me a possible future in which powerful "replicators" are widely available. To purchase a car in such a world, you would insert your credit card into a machine containing the digital designs for hundreds of cars. You would select the car you desired, and the machine would use that car's digital blueprint to automatically manufacture the car for you on the spot. In this scenario, Ford Motor Company might never ship a physical car anywhere; instead, the company would only distribute digital instructions for building cars.[36] Legal standards attempting to count the number of times a product has been supplied from one location to another would stumble in such a world, in which products would effectively be teleported from manufacturer to consumer rather than shipped in any traditional sense.

Such blurring of the distinction between descriptions of machines and the machines themselves will make it increasingly difficult to prohibit patents on scientific principles but allow patents on practical applications of those principles. Traditionally, when someone discovered a new scientific principle, the time and effort required to apply the principle in practice (such as by inventing a machine based on the principle) in effect created some breathing room for the principle to spread and be absorbed by the public, free of any patents attached to it. Even when people did obtain patents on machines that applied a scientific principle, such patents tended to be narrow in scope—such as Morse's telegraph patent implementing the principle of electronic transmission of information— because of the difficulty of quickly inventing a wide range of devices applying the principle. Real-world friction of this kind, which slowed the transformation of newly discovered unpatentable principles into patentable machines, is being eliminated by the lubricating effect of artificial genies. In the Artificial Invention Age, it will tend to be easier for those who discover new scientific principles and other abstract ideas to produce designs for broad ranges of inventions based on those ideas and patent them early, potentially enabling control over practical uses of such ideas to be concentrated in the hands of a few before such ideas have time to percolate into the public body of scientific knowledge.

More fundamentally, the traditional basis for granting patents on practical inventions and not on abstract scientific knowledge is the high cost and risk involved in transforming the latter into the former. If artificial invention technology is reducing such cost and risk, it is eroding the justification for patents. I tend to think improvements in artificial invention technology will not necessarily reduce the *overall* cost and risk of inventing, but rather just *shift* the cost and risk of writing wishes up to higher and higher levels of abstraction, without limit. I could be wrong, however; only time will tell.

The final dividing line that artificial genies will whittle away is the one between the liberal arts and the useful arts. For example, the techniques used by artists and engineers to create works in their respective crafts will overlap and merge. We've already seen the similarity between the process of designing a microprocessor by writing an HDL description of it and the process of writing a hierarchical outline for a book. Computer programmers have long mixed and matched program-writing strategies from engineering and creative writing. This melding of creative *techniques* across the two cultures will be mirrored in wishes themselves, which are at once a kind of literary work and the most practical of engineering artifacts. We'll see in Chapter 14 how this is already creating a class of "renaissance geeks," who not only know how to write computer code but understand the genetic code, who know how to solve problems themselves and manage teams of problem solvers, and who read circuit diagrams as well as balance sheets. This is nothing less than the death of our species' current antlike specialization[37] and the return of the generalist artisan—a welcome return in my view. The bright-line distinction between the liberal arts and the useful arts is relatively recent in historical terms in any case,[38] and it has probably reached the end of its useful life.

The traditional inverted-Y funnel shape that steers creations in the liberal arts through copyright law and those in the useful arts through patent law will turn into a sieve in the Artificial Invention Age. Our current system, in which software straddles both copyright and patent law, is already exhibiting some bizarre side effects, such as the fact that the ability to patent written computer source code creates patents that could be used to restrict speech.[39]

Furthermore, recent court decisions, reflecting an inability to toe the line between the liberal arts and the useful arts, have allowed business methods to be patented. Traditionally you could not patent a performance-based scheme for compensating the manager of a private company. Yet the U.S. Patent

Office's Board of Appeals reversed the rejection of a patent application for just such a scheme.[40] Opening the door to patents on business methods has also encouraged enterprising inventors to submit patent applications for a host of methods falling within the liberal arts and not involving the use of any machinery whatsoever:

- Obtaining a percentage of total profits from exclusive marketing arrangements with independent producers of different products
- Teaching a vocabulary word by giving a student a recording of a song containing the word
- Facilitating conversation between at least two dining partners by giving them printed open-ended questions to ask each other
- Creating an expression of a positive feeling such as love toward another person by placing a note expressing the positive feelings in a package and delivering the package to the person with specified instructions[41]

If you think we needn't worry about these examples because they have not yet been granted as patents, consider that a U.S. patent has been granted on a method of sharing erotic experiences by providing a building with a purportedly novel kind of peep show.[42] We've gone from patenting the steam engine to patenting steamy experiences.[43]

Although commentators love to blame the U.S. Federal Circuit Court of Appeals for opening patent law to business methods, the Court is only acting in response to forces greater than it and beyond its control. Computers are automating business methods and other techniques in the liberal arts at breakneck speed. Law firms are already servicing their clients with Web-based software that will answer their questions about tax law without the need to consult a human attorney.[44] Software is writing poetry, teaching children in classrooms, recognizing and translating speech, and holding online auctions—all functions that previously could only be performed by humans working in the liberal arts.

By increasing the ease with which business methods and other methods in the liberal arts can be automated using computers, artificial invention technology is whittling away at bright-line legal rules, such as "no patents for business methods," which courts have relied on to make it easier to decide individual cases. We shouldn't be surprised that once such cracks appear in the dam, enterprising patent applicants will materialize to test the limits of the remaining edifice.

Although most of us would agree that techniques in the liberal arts that are performed "by hand"—such as a new way of painting with a brush—should not be patentable, once such a method is "bottled" inside software or a machine, the question is much closer. Any flat prohibition on patents for machines that perform business methods would exclude much that has been patentable for ages. Perhaps the best proof that machines performing business methods cannot be *per se* unpatentable is that the world's largest patent holder, International Business Machines (IBM), has "business machines" right in its very name.

I suspect that at some point the accumulation of these cracks in the dam will require nothing less than a new system of intellectual property. Many such systems were proposed in the 1980s and 1990s in response to the problems raised by software for traditional intellectual property law. One of the more notable recommendations was in *A Manifesto Concerning Legal Protection of Computer Programs*, published in 1994 by a team of legal and technical experts, headed by law professor Pamela Samuelson at the conclusion of a multiyear effort funded by the Kapor Family Foundation.[45] The *Manifesto's* basic conclusion was that software should be protected by a market-oriented approach whose goal is to give software developers a reasonable lead time (the time *after* they introduce their software to market and *before* they face competition in the market). The *Manifesto* calls this an "unobstructed opportunity to seek market reward before imitations can lawfully appear there." For such a system to be attuned to the features of various software markets, it would need to "be tuned to the 'basal metabolic rate' of the market"—in other words, the rate at which new products are developed and brought to market.[46] Software that could be developed more quickly would generally be protected for shorter periods of time. As a result, software in a number of markets might be protected for varying lengths of time, unlike the current "one term fits all" system.

The *Manifesto* was an admirable effort, particularly in the way that it cut to the chase by proposing a solution that directly addressed the professed purpose of intellectual property law—to provide incentives to produce and disseminate works to the public by offering reasonable assurance of return on investment—in a way that was sensitive to patterns of innovation in numerous industries at different times. Despite its strengths, the *Manifesto* is already quickly growing outdated in light of artificial invention technology. Its recommendations make sense in a world in which biotechnology is developed by a biotechnology industry having a certain identifiable mode of inventing associated with certain kinds of cost (e.g., on the order of $1 billion per new drug), in which automotive technology is developed by an automotive industry having a certain identifiable way of inventing associated with particular kinds of cost, and so on. In such a world, intellectual property rights can reasonably be tailored to technology industries.

As all of these old distinctions continue to blur, and as artificial genies automate the process of transforming concepts from one domain into another—equation into car frame, words into machine—attempts to rely on the old distinctions will become increasingly less reliable and the consequences of miscategorization more severe.[47] As an example, consider that in Morse's day the cost of granting a patent that should have been denied was relatively small because patents were relatively narrow. If one slipped through the cracks now and then, its ability to impede innovation by others was minimal. But in the Artificial Invention Age the cost of wrongly granting a patent on a wish for a broad category of new technology could be devastating, by enabling its owner to block innovation in an entire field. Yet the cost of wrongly deciding to *reject* such a patent could be equally devastating, by driving wish writers to keep their wishes secret or stop writing them at all. Small errors in such decision making could have significant consequences. It's like the difference between giving your five-year-old a peashooter or an Uzi with which to shoot a poisonous snake off your arm. This is why I have repeatedly emphasized the need for the legal rules and practices I suggest to be applied particularly rigorously if they are to have their intended consequences.

Breaking down barriers is exciting, but it also creates uncertainty. Although new categories will arise, we can't yet know what they will be. The best we can do in the face of such uncertainty is to stay tuned and attuned to the changes with eyes wide open so that we can respond to them nimbly and appropriately. Maybe we will even be able to enlist some genies to help us find the right solutions.

III LIVING WITH GENIES

11 Learning How to Wish

NOW THAT WE HAVE LEARNED how artificial invention technology works and how patent law should adapt to it, we turn to the implications of invention automation for four representative groups of people—inventors, high-tech businesspeople, legal professionals, and consumers—and the strategies that each should adopt to prosper in the Artificial Invention Age.

If you are an inventor, and artificial invention technology is automating your most coveted skills, the good news is that you will not necessarily become obsolete.[1] In fact, if you play your cards right, you will be able to leverage artificial invention technology as a kind of "brain booster" to augment your inventive skills and thereby make you a better, more valuable, inventor. The bad, or at least challenging, news is that to do so successfully will require you to update your existing skills and develop new ones. Fortunately, the early stages of the Artificial Invention Age offer clues about what skills you will need so that you can begin to develop them now.

An Artificial Invention Skillset

Abstract Problem Definition

> By relieving the brain of all unnecessary work, a good [mathematical] notation sets it free to concentrate on more advanced problems, and in effect increases the mental power of the [human] race.
>
> —*Alfred North Whitehead*

Because computers will automate many low-level skills required of today's inventors—such as designing the shape of an antenna or writing code for software—savvy inventors will focus their energy on developing skill at high-level, abstract problem solving instead. The first, and most direct, consequence of this increasing demand for abstract problem-solving skill is that tomorrow's inventors will need to learn how to write problem descriptions in the language of artificial invention technology. This will put them in a position that more closely resembles a CEO than a traditional engineer. Henry Ford didn't design the revolutionary V-8 engine himself. Instead, he told his engineers to create "an engine with all eight cylinders cast in one block" and then left it to them to work out the details. Although at first they told Ford that such an engine was impossible, they eventually succeeded in designing the V-8 engine, and as they say, the rest is history.[2]

As even this simple example illustrates, it would be a mistake to conclude that the work of inventors will necessarily become *easier* as it becomes more abstract.[3] Ford required an ability to think outside of the conventional wisdom of the day to formulate the abstract but concise directive he issued. Developing such a vision and expressing it clearly is not necessarily easier than designing an automobile engine itself, just different.

To be sure, defining a problem abstractly *can* be easier than working out the problem's solution in detail. Writing the computer program "Add 2 + 4" is easier than designing circuitry from scratch for calculating the same sum. Furthermore, the task of formulating a high-level problem statement certainly is easier than performing the *entire* process of formulating such a statement *and* solving the problem yourself.

This does not mean, however, that developing a clear and concise statement of a problem in an abstract language that a computer can understand— what I have been calling writing a wish—is necessarily easier than the task of solving the problem directly.[4] Inventors invent mousetraps without understanding or being able to explain why they work in the abstract. Mathematics offers the clearest proof that abstract problem solving is not necessarily easier than low-level problem solving: problems in pure mathematics are among the most abstract *and* the most vexing that we face. Solving problems directly at a low level of detail can be easier than solving the problem abstractly if you don't understand the mathematical, scientific, or other theory underlying potential solutions to the problem.

Another reason the work of inventors won't necessarily get any easier in the Artificial Invention Age is that inventors, with access to software that puts the equivalent of Henry Ford's team of engineers at their disposal, will quickly use the technology to pick the low-hanging inventive fruit and turn their attention to solving problems at the cutting edge of difficulty at the next highest level of abstraction. Recall from Part I that this is exactly the progression we have witnessed in the history of computing. Once computers made it possible for programmers to write programs in machine language rather than by manually rewiring circuitry, programmers quickly wrote increasingly powerful and complex programs in machine language that tested the language's limits. What had started as a tool for *simplifying* problem solving enabled programmers to move on to tackling more *difficult* problems using the tool. The solution? To create high-level programming languages, which enabled programmers to write increasingly abstract and powerful programs until the limits of those languages were reached, and the cycle continued. The same feedback loop between inventors and the invention-automating tools they create will ensure that wish writing remains challenging throughout the Artificial Invention Age.

Furthermore, in practice the best wish is not necessarily the most abstract one. Recall that, thanks to the combinatorial explosion, it is impractical for software running on even the most powerful computers to search through more than a tiny fraction of the potential solutions to any given problem. Therefore, inventors in the Artificial Invention Age will often need to develop the intuition and skill necessary to front-load their wishes with clues about how to solve the problem at hand.[5]

Moreover, not all wishes take the form of explicitly stated abstract criteria. As we saw in the CrossAction toothbrush example, sometimes a wish takes the form of a set of *data*—such as designs of existing toothbrushes and tooth-brushing performance data—that may not yield any clear indication to a human of the criteria that a problem solution must satisfy. Instead, the computer's task is to attempt to extract such criteria from the data themselves and to use that extracted knowledge to forge a solution. Inventors who use this kind of artificial invention technology will require skills such as the ability to determine which kind of data to gather, how to obtain the data, and how to represent that data in the form that is most efficient for a computer to process.

Inventing Genies

There also will be an increasing need for inventors who can invent better artificial genies. Many of the inventors I interviewed for this book did not simply use off-the-shelf software to produce the impressive results you've seen; they modified an existing genetic algorithm, neural network, or other process to optimize it for solving the problem at hand. Inventors who have the ability not only to write wishes but to custom-fit a genie to match their own needs will have a leg up in the Artificial Invention Age.

As we saw in Part I, many kinds of artificial invention technology rely on software to simulate the operation of potential designs—such as the flow of air around a car frame—to determine whether they can solve a given problem. The more accurate and faster the simulator, the better and more quickly the artificial invention technology can produce results. As a result, today's artificial inventors are always on the lookout for better simulators. Therefore, we can expect people with expertise in physics and computer programming who can create better simulators to be in high demand in the Artificial Invention Age.[6]

This is just one example of how expertise in pure mathematics and physics will become increasingly valuable in the Artificial Invention Age. Every math major who was ever told to study a more "practical" subject, such as mechanical engineering or chemistry, will stand vindicated. Skill in these subjects will be valuable not only for creating better simulators but also for writing wishes in the abstract language of mathematics and physics that artificial genies demand.[7]

Automation of particular skills does not necessarily eliminate the jobs of people who use those skills in their jobs. Instead, automation can shift the division of labor between machine and human, thereby redefining the meaning of a given job title. Although today we use the term *automotive engineer* to refer to someone who designs the specific components of an automobile engine, tomorrow's automotive engineer will be someone skilled in mathematics and physics who can write the high-level requirements that cutting-edge automobile engines must satisfy.

We already experienced this shift particularly acutely as the meaning of "computer programmer" morphed from someone in the 1950s who manually rewired a computer plugboard to someone today who sits at a keyboard writing instructions in an abstract programming language. Same job title, different skills. Go back far enough in the history of computing and you can no longer find people called "computer programmers," just electrical engineers who both designed and programmed computers. This shift over time in what it means to be a programmer has not necessarily made programmers as a class obsolete; only those programmers who could not update their skills to keep pace with advances in the automation of programming disappeared.

Biology

Traditionally, computer programmers have been trained either directly in computer science or in some combination of computer science and electrical engineering. In recent years, an increasing number of programmers have fashioned successful careers solely on the basis of knowledge of programming, without any understanding of the hardware that makes their creations possible. Very rarely do you find a computer programmer with expertise in both computer science and biology.

This situation presents a valuable opportunity for inventors in the Artificial Invention Age, because knowledge of both computer science and biology is particularly useful for developing *and* using many kinds of artificial invention technology. Two of the most common technologies—evolutionary algorithms and neural networks—are modeled on biological processes and systems (namely, biological evolution and the human brain).[8] Although early incarnations of these technologies mimicked real-world biological systems only crudely, more recent work has striven to make digital biology imitate its real-world counterparts more accurately in an attempt to emulate the ability of natural systems to solve complex problems. The success of these efforts will depend on the ability of tomorrow's inventors to understand the natural world and translate their understanding into improved artificial invention technology.

A firm grasp of biology will be critical to those who develop artificial invention technology, and to those who endeavor to use it to its maximum benefit. Real-world biological evolution is constrained by the forces of nature (such as gravity) and by the natural raw materials (such as carbon, hydrogen, and oxygen) available to it to generate organisms. Inventors who use genetic algorithms have the liberty to set up the artificial equivalents of these initial conditions and other natural laws of simulated evolution, such as the rate at which mutations occur and how often organisms mate in each generation. Although you could use a genetic algorithm to invent merely by giving it a wish (fitness function) and then letting it run free with any set of constraints, choosing the constraints carefully can significantly improve the end results. Similarly, tweaking your wish according to knowledge of the simulated environment in which it will be applied can increase the likelihood that the wish will be granted.

Business and People Skills

Learning to think more like Henry Ford will mean thinking more abstractly and also developing management, interpersonal, and other business skills.

Freeing inventors from the need to focus on low-level technical details will create an opportunity and a demand for inventors who can see the bigger picture in all of its facets—a skill in short supply among today's inventors. The archetypal engineer of today, who spends his days alone in a cubicle, focused single-mindedly on solving a narrow technical problem, will face extinction, because once technical ability can be replicated more quickly, less expensively, or more quickly by someone else—whether the someone is a human or a machine—one's only chance at success is to compete on the basis of higher-level (or otherwise specialized) skills that have not yet become automated or outsourceable less expensively.

The skills required to be a successful "information technology" worker are already shifting in response to the pressures being put on workers in the United States and other high-tech centers by outsourcing to countries such as India and China for low-level programming jobs. Such outsourcing leaves behind a demand in the United States and other outsourcing countries for business analysts and IT relationship managers. People with experience in specific industries—"domain experts"—will be in particularly high demand.[9] Those with the ability to learn about new domains quickly will be the most valued of all, because such an inventor will be able to solve a problem in the automotive industry one day, the financial services industry the next day, and the health care industry the day after that.[10] Such domain hoppers will use the knowledge they gain from seemingly disparate fields to produce solutions that would not have been possible to create by applying only narrow engineering expertise in a particular domain, using what Dan Pink calls "symphony": "the capacity to see relationships between seemingly unrelated fields; to detect broad patterns rather than to deliver specific answers; and to invent something new by combining elements nobody else thought to pair."[11]

This is just as true for businesses as it is for individuals. We should expect, for example, that in the latter stages of the Artificial Invention Age inventive skill itself will become a commodity and thereby lose its ability to confer a competitive advantage on businesses. For a clue as to where companies might turn next for that advantage, we need only look to what makes some existing software development projects more efficient than others. A recent study found that "the best in class software development projects are 3.37 times faster to market and 7.48 times cheaper than the worst." Management and technology *approaches*, not technology itself, constituted three of the four factors that contributed to these results. In other words, the best software development projects run as smoothly—and hence quickly and inexpensively—as they do primarily because of how people manage and execute those projects. Even in a field that has been driven so much by computer automation, there is and likely will continue to be substantial room for human expertise to make a significant real-world difference. Such expertise, however, will lie as much in management skills as in technology skills. Dilbert's long-derided "pointy-headed manager" may yet have his day in the sun.

For example, companies in need of improved technology do not always know ahead of time what kind of technology best suits their needs, something that Jeff Walker of TenFold refers to as the "requirements gap." TenFold attempts to narrow this gap by using what Walker calls "extreme rapid application development." TenFold's customer uses the company's software to describe the customer's initial idea of its requirements and quickly build an initial version of a software application that satisfies those requirements. The customer then immediately begins using the software prototype. Through this experience, the customer may change the initial idea of requirements, for example by realizing that it needs stronger security but not the advanced accounting features it originally thought were critical. The customer can then use the updated requirements to quickly produce a new version of the software application, and repeat as necessary until the requirements converge with the software it produces.[12] Such an iterative development process is feasible only if new versions of the software can be described and generated quickly. TenFold addresses the needs of customers not merely by offering a technically advanced solution but by tailoring the solution to how customers think and work.

Even when a company does know what problem it is trying to solve, it may not be able to *quantify* that need. A bank may recognize it would benefit from understanding which kinds of loans on its books are likely to default in the next quarter but lack a way to categorize the relevant features of its existing loans with the mathematical precision necessary to enable such predictions to be made.[13] The ability to help others quantify such needs in a form that a computer can process will be highly valued in the Artificial Invention Age. Even if the bank is fortunate enough to have access to off-the-shelf artificial invention software, and to have an employee with basic knowledge of how to write wishes for the software, those with broader and deeper wish-writing experience will still be in demand because "if you work in the area long enough, you start to get some of the insights about what works and what might not work, and . . . a company that has read one book on genetic algorithms would benefit from hiring someone that's worked in the field for many, many years."[14]

The Ability to Adapt to Skills Not Yet Automated

Whether or not all of my predictions about the skills that tomorrow's inventors will need are accurate, what ties all of these predictions together is the increasing necessity for inventors to develop interdisciplinary skills and the ability to adapt flexibly to future automation by acquiring skills that cannot be performed automatically more efficiently than manually. The inventor of the future will blow past artificial divisions between fields of knowledge and instead draw on knowledge from any field that is useful for solving the problem at hand.

As a recent *New York Times* article put it, what we need are not traditional code writers but renaissance geeks. The article cites a computer science Ph.D. candidate at Virginia Tech whose research includes work in anthropology, sociology, psychology, and psycholinguistics. "For students like Ms. Burge," the article continues, "expanding their expertise beyond computer programming is crucial to future job security as advances in the Internet and low-cost computers make it easier to shift some technology jobs to nations with well-educated engineers and lower wages, like India and China." Many such students finish their studies to embark on careers in medicine, law, media, the arts, and scientific fields other than computer science.[15]

Although one might attempt to maintain job security in such an environment in other ways, the alternatives are at best riskier and at worst doomed to fail. One might take the approach still championed by the legal profession, namely to erect legal barriers to competition (such as by requiring a professional license to be an inventor) and corresponding economic barriers (such as the high cost of obtaining such a license). But the medieval guild approach is rapidly breaking down even for lawyers—its most stalwart defenders—in the face of increasingly global competition and automation.

Alternatively, one might try to predict which skills will elude automation and outsourcing for a long enough time and then sell products and services using those skills. But predicting the future with this degree of specificity is a tricky business.[16] As Niels Bohr said, "Prediction is very difficult, especially about the future." Just ask a travel agent, encyclopedia publisher, or anyone in the newspaper or music industry how successfully they predicted the impact of computer technology on their businesses even five years ago. I recommend placing your bets on the skills necessary to gain new skills, rather than on your ability to predict whether any particular skill will resist automation or outsourcing.

In his excellent book *A Whole New Mind*, Daniel Pink argues that in what he calls the "conceptual age" people will benefit from developing their right brain (holistic) skills rather than their left brain (reductionist, analytical) skills. Pink, though otherwise observant and insightful, misses the ways in which elements of both left-brain and right-brain thinking will be useful to artificial inventors. There is nothing more left-brain than skill in mathematics and physics, and nothing more right-brain than big-picture, abstract problem solving; yet artificial inventors will need a healthy dose of both. What will define the Artificial Invention Age, and the skills needed to thrive in it, is not the distinction between left brain and right brain, between the liberal arts and the useful arts, between science and engineering, between philosophy and science, or between any other preexisting dichotomy. Rather, it will be the distinction between *defining the problem*—which we will still need humans to perform—and *specifying the solution to the problem*—which will become the province of artificial invention technology.

Even understanding that this is the right distinction to draw does not make it particularly easy to predict where the dividing line will fall at any particular point in the future, because what constitutes defining the problem and what constitutes specifying the solution to the problem are both relative and dynamic. Today's human-written wish is tomorrow's computer-generated solution. Machine-language programmers enjoyed job security in an era when only humans could write machine-language code. Then compilers automated the transformation of high-level instructions into machine-language code and the demand for machine-language programming skills decreased. As artificial invention technology automates inventing at higher and higher levels of abstraction, the previous high levels of abstraction at which human skill is required to write problem definitions become low-level problem solutions no longer requiring human skill to generate. As a result, no particular skill is guaranteed to qualify as skill in problem definition in absolute terms and for all time. Therefore, although humans will always be needed to perform problem definition (make wishes), for better or worse the meaning of "problem definition" will change over time as a function of the current state of invention automation.

Artificial inventors will best leverage their skills in many situations by combining old-fashioned inventing with automated inventing. Because technologies such as genetic algorithms and neural networks excel at quickly finding "pretty good" solutions that overcome the blind spots of human inventors, such technologies can be used to develop an initial prototype, which human inventors can then refine using traditional engineering techniques.[17] Conversely, one might produce a prototype using traditional engineering techniques and then refine it using artificial invention technology.[18] Savvy inventors who mix and match in this way will not only be able to produce better inventions more quickly than would be possible using artificial invention technology or old-school inventing in isolation; they will also be able to *learn* from the results of artificial invention technology and apply their new knowledge to future inventive endeavors. Humble inventors who recognize their own limitations will, ironically, be precisely those who will be able to learn from artificial invention technology and move beyond those limitations in ways that would not have been possible had they relied on human ingenuity alone.

Peril and Promise for Inventors

Peril

> I have created a machine in the image of a man, that never tires or makes a mistake. Now we have no further use for living workers.
> —*Rotwang, in Fritz Lang's Metropolis*[19]

Inventors who confront the picture I have painted may experience one of two extreme feelings: exhilaration or dread. Both are legitimate and well founded. Revolutionary times are trying times, even for those who stand to benefit from them, particularly when success is not guaranteed.

We don't have to look into a crystal ball to see evidence that justifies today's concerns among inventors about the futures of their careers. Technological obsolescence, automation, and outsourcing have all put pressures on engineers in recent years to upgrade their skills or join the ranks of the unemployed. Although some continue to maintain that *inventive* skills can never be automated, arguments that particular skills are impervious to automation do not have a successful historical track record; just ask your neighborhood blacksmith, millwright, or tanner if their skills will withstand automation. The link between automation and human obsolescence can be both direct and dramatic; legend has it that Jacquard, the inventor of the programmable loom that we encountered in Part I, was attacked in the streets by "draw boys" whose jobs had been made obsolete by his handiwork.[20]

Many respond to the concerns of today's inventors in the United States over outsourcing by arguing that it will benefit such inventors by relegating low-level, and therefore lower-paying and less-interesting, work to engineers in other countries, thereby freeing U.S. inventors to engage in more abstract and intellectually satisfying work as well as earn higher wages to boot. Although this is one possible outcome, and the one on which I focus primarily in this book, it is not preordained. Whether or not the future is bright for U.S. inventors depends on a combination of economic, legal, political, and social forces whose interactions are difficult to predict.

Furthermore, even though technologists in India, China, and other countries currently benefit from outsourcing, there is nothing to stop their work from being outsourced or automated in the future. In fact, the term *reverse outsourcing* has already been coined to refer to the employment of workers in the United States and other Western countries by outsourcing firms in India and elsewhere as a way to satisfy the Western demand for outsourced services.[21] We can only expect that if invention automation yields greater cost savings than outsourcing in particular circumstances, market forces will shift demand toward invention automation technology and away from outsourced human labor in those circumstances. Inventors in countries that offer outsourcing services, therefore, should pay as much attention to the potential impact of the Artificial Invention Age on them as inventors in the West.

Even if invention automation technology does not put human inventors out of work, it could "deskill" the work human inventors perform. Software guru Robert L. Glass has argued that although milestones in the history of software engineering—such as the development of high-order programming languages, operating systems, and modular programming—reflected great creativity on the part of those computer scientists who *invented* the milestones themselves, it is at least worth asking whether those milestones contributed to or detracted from the creativity of *future* programmers who made use of such milestones. As one example, Glass cites the work of two authors who argued that the effect (and possibly even intent) of the managers who introduced methodologies such as structured programming and object-oriented programming into the workplace was to constrain how programmers could write programs, thereby deskilling them rather than enhancing their creativity.[22]

This is merely one example of how improvements in technology or methodologies developed by some inventors can harm the position of others. It also illustrates well how more abstract or high-level thinking is not necessarily more interesting or nourishing of creativity than concrete or low-level thinking. Computer technology may automate problem solving to such a degree that the problem-defining work remaining for humans to perform is both eminently abstract and stultifyingly boring.[23] In this case, inventors will be made subservient to machines, as in the future postulated by Samuel Butler in which "man shall become to the machines what the horse and dog are to us."[24]

Although the downside potential for inventors is real, nothing can be done to close Pandora's box. If even a single inventor writes software that can automate a task previously performed by human inventors and releases the software to the public, there is nothing to stop anyone else from using it instead of a human inventor to perform the same task. Therefore, the best inventors can do is enter the Artificial Invention Age with their eyes wide open, fully aware of the risks and of the steps they can take to minimize those risks and maximize the gains to themselves.

Promise

The best possible outcome for inventors from the new division of labor between human and machine is that human and computer will each do what it does best, thereby freeing humans to be more human.[25] This potential is borne out by the early successes of automated calculating machines, which both reflected and helped to reinforce a growing belief that people should not be spending their days crunching numbers; doing so was "turning men into veritable machines."[26] It has even been said that "the reason for building a personal computer in the first place was to enable people to do what people do best by using machines to do what machines do best."[27]

Computers, for example, may never be able to make aesthetic judgments, such as deciding which of two perfumes smells sweeter. This doesn't mean, however, that artificial invention technology can't be used to invent new perfumes, only that a human's sense of smell and aesthetic judgment are required to tell the computer which perfumes are better than others. The human's computer partner can do what it does best, namely use the human judges' input to further refine the potential perfumes that it produces in the next round for human evaluation. This iterative feedback loop makes maximal use of the strengths of both human and computer. Here are just a few examples of this kind of collaborative human-computer tag team:

- Luis von Ahn at Carnegie Mellon University is using online games—such as the ESP Game (www.espgame.org) and Peekaboom (www.peekaboom

.org)—to harness the power of Internet users worldwide to label digital images, a task that is nearly impossible for software to perform, thereby improving the ability of von Ahn's software to search for and recognize images.[28]

- Karl Sims used "interactive evolutionary computation" to produce a media installation named Galápagos that allowed visitors to "evolve" 3D animated images of virtual "organisms." Twelve computers with displays were lined up side-by-side, displaying an initial population of organisms. Viewers selected the organisms they liked the most by standing on sensors in front of those organisms' displays, causing those organisms to "mate" with each other and produce offspring. Visitors then watched the evolution take its course on-screen in response to their selections.[29]

- Jason Calacanis recently started an Internet search engine named Mahalo ("thank you" in Hawaiian) that is powered not directly by software but by *humans* who spend their days searching the Internet for Websites on the most popular topics ("Paris Hilton" and "iPod" at the time of this writing) and writing up concise lists of useful links to those topics. Mahalo attempts to combine the best of human skill *and* search technology; Mahalo's human searchers use Google and other existing search engines to find their source material.[30] By leveraging this combination of carbon and silicon, Calacanis is hoping to best Google's 10,000 employees with just 60 of his own.

Each example demonstrates how a partnership between human and computer can achieve collectively what neither can accomplish individually by combining their strengths. Savvy inventors who realize this in the Artificial Invention Age will recognize the limitations of the technologies they use, so that they can enlist human expertise at just the spots where computers choke.

The impact of computer technology on *non*inventive jobs has long been recognized. For example, Frank Levy of MIT and Richard J. Murnane of Harvard aptly describe in their 2004 book, *The New Division of Labor: How Computers Are Creating the Next Job Market*, how computers have radically altered the tasks performed by people ranging from bond traders to cardiologists to mortgage brokers to secretaries. The last of these is telling. The professors point out that the U.S. Department of Labor *Occupational Outlook* of 25 years ago defined the role of a secretary as "reliev[ing] their employers of routine duties so they can work on more important matters" and listed the specific tasks performed by secretaries to achieve this end as typing, taking shorthand, and dealing with callers. The most recent version of *Occupational Outlook* pointed out that "office automation and organizational restructuring have led secretaries to assume a wide range of new responsibilities once reserved for managerial and professional staff," including "training and orientation to new staff, conduct[ing] research on the Internet, and learn[ing] to operate new office technologies."[31]

Computer automation, in other words, has not necessarily *eliminated* the need for secretaries; it has reassigned *higher-level* tasks to secretaries in service of fulfilling their original calling of "reliev[ing] their employers of routine duties so they can work on more important matters." As computers have automated an increasing variety of tasks previously performed by human secretaries, the tasks performed by secretaries have shifted to those that computers and other technologies *cannot* perform (yet). Although Levy and Murnane focus on jobs that traditionally have not involved *inventing*, the same general conclusions apply to the changing division of labor between inventors and computers.

If it is true that the role of human inventors will be redefined over time as "whatever tasks in the inventive process need to be performed but that computers currently cannot do," then we would do well to predict the role of human *inventors* at any point in time by identifying the limitations of artificial invention software at that time. If artificial invention technology cannot automate some part of the inventive process, then human inventors will still be needed to perform that part, even if computers do everything else. Inventors who wish to keep their skills as up-to-date and marketable as possible, therefore, will do well to focus sharply on the division of labor between human and computer, as defined by the changing capabilities and limitations of both. Those who do so will both maintain job security and find they can leverage computer technology to increase their inventive powers to previously undreamed-of heights.[32]

12 Competing with Genies

MODERN INVENTORS OFTEN INVENT not within their own private laboratory but within a large organization such as a private corporation or a government agency. Typically such entities, not the individual inventors, transform inventions into marketable products and services. In the Artificial Invention Age, even a company with the world's best inventors, expertly trained at using the latest artificial invention technology, will fall to its competitors if it does not adapt its business strategies to the new environment. We now explore the concrete steps that high-tech businesses must take to obtain maximum advantage from artificial invention technology.

Hire Expert Wishers

The first and simplest step that businesses should take to capitalize on artificial invention technology is to hire expert wishers—inventors who are skilled in the use of artificial invention technology in the ways described in Chapter 11. This does not necessarily mean hiring such inventors as employees; as we will soon see, companies need to choose carefully between developing or using artificial invention technology in-house on one hand and outsourcing the use of that technology on the other. Furthermore, businesses must find inventors or external firms who have not only the right technical skills but also a knowledge of the business's industry and the ability and willingness to work with the company as a close partner to solve the business's current needs. The renaissance geeks we met in Chapter 11 will fit right into this business plan.

Rethink the Costs of Invention

Using inventors skilled in artificial invention technology will enable businesses to reduce their labor costs and overall research and development costs because each inventor's productivity will be boosted by artificial invention technology. I've been told that even today a college graduate using evolutionary computation software can often match the inventive capabilities of someone with a Ph.D. who invents "manually."

More generally, artificial invention technology will radically change the cost structure of high-tech companies. Already, artificial invention consultancies such as Imagination Engines, NuTech Solutions, Genalytics, Matrix Advanced Solutions, and Icosystem are using artificial invention software to enable other companies to design higher-quality products more quickly and inexpensively than is possible using human inventors alone. Here are just a few real-world success stories, drawn from a range of industries, in which artificial invention technology has been used to reduce the amount of time or money required to solve a problem:

- Natural Selection, a firm specializing in the use of artificial invention technology, used evolved neural networks to replace toxicological tests for lead-based drugs, which previously took two years, with a new neural network-based test that takes only days to complete.[1]

- i2 Technologies has used genetic algorithms for supply chain management to produce anticipated savings of $1 billion for Boeing, create a 500 percent return on investment (ROI) for Dell, reduce Barnes & Noble's inventory by $13 million, and cut Whirlpool's inventory by $4.8 million.[2]

- NuTech Solutions used a combination of genetic algorithms, neural networks, simulated annealing, evolutionary computation, and swarm intelligence to produce a "global optimizer" for General Motors that reduced the time of one design process from five weeks to two and reduced the weight of an automobile frame by 30 pounds.[3]

- Imagination Engines used the Creativity Machine to invent 200 new supermagnetic rare-earth iron boride composites, each with comparable magnetic properties to neodymium-based borides but at nearly half their price.[4]

- The software that we encountered from TenFold in Part I can generate corporate software applications in three weeks, compared to more than a year when done by hand.[5]

With artificial invention technology increasing the speed and reducing the cost of inventing without sacrificing quality, high-tech companies will have to further rethink many of the assumptions they have already begun to reexamine in light of outsourcing. To what extent should the company retain full-time employees whose sole function is to invent, in contrast to outsourcing the work of invention to firms that specialize in using artificial invention technology to invent? Will the technology enable product development cycles, which in many cases have already shrunk from years to months, to shrink further to months or even weeks? If so, what impact will this have on the company's marketing strategy and other aspects of its business model? If a company that historically has focused on developing new products and selling them to customers now develops an expertise in using artificial invention technology to invent, should it shift or expand its business model to include provision of inventing services to other companies lacking the ability to use artificial invention technology efficiently? To the extent that the technology increases the ease and reduces the cost of inventing not only to the company's competitors but also to its customers, what core competitive advantage will remain for the company to exploit?

At this early stage in the Artificial Invention Age, I can only pose these difficult questions rather than give answers. The full impact of the reduced cost of inventing will inevitably be complex and change over time. For this reason, however, every company will need to stay on its toes and remain poised to adapt to the shifting cost of invention as it evolves, particularly because throwing artificial invention technology into the mix can produce counterintuitive results. For example, it may be economically prudent for a company to use a genetic algorithm to design a product even if the algorithm produces a design that is *less* understandable and of *lower* quality than could be produced in theory by a human designer, if the cost of using the genetic algorithm to produce the design is low enough. Companies are likely to encounter such tradeoffs with increasing frequency as the cost of expert human labor rises and the cost of computer technology falls.[6]

On the other hand, companies relying too heavily on using the least expensive technology and outsourcing to the lowest bidder for every new product risk missing out on the deep understanding of a company's products and ser-vices that only people (whether employees or contractors) who have long-term relationships with the company can possess.[7] Such interlocking incentives may result in an economy built on both small companies—as small as one employee—and large companies that "have almost as many 'centers' as they have people."[8] Thomas Malone at MIT argues that the decreasing cost of *com-munication* technology has enabled small companies to experience the best of both worlds: the economies of scale and other economic benefits of large com-panies, combined with the human benefits of small companies, such as flex-ibility and creativity.[9] The decreasing cost of invention augmentation technol-ogy promises to do the same for inventors and the companies for which they work, potentially enabling even individual inventors and small engineering shops—what Dan Pink calls "microbusinesses"[10]—to compete head-to-head against much larger technology research and development firms.

Think Mix and Match, Not Either-Or

The Artificial Invention Age will present forward-looking businesses with more than just a binary choice between old-fashioned inventing and artificial inventing. It is true that in the simplest case a business currently using human engineers to invent in 19th-century style will stand to benefit from giving those engineers artificial invention technology, training them to use it, and keeping everything else in the business unchanged.

But the real benefits will come from using artificial invention technology in those situations and in those ways in which it is most effective, and in com-bining new kinds of inventive technology with old-fashioned human exper-tise and innovative business models. In recent years, outsourcing has become all the rage for reducing a company's costs by using lower-cost contract labor, typically in another country, to provide services to the company that pre-viously were performed by its own employees.[11] Although at first outsourc-ing was used exclusively for administrative or otherwise "noncreative" tasks, more recently companies have begun to outsource low-level programming and other "creative" jobs.[12]

A company looking to reduce the cost of performing a certain business function faces the choice between outsourcing the function and improving the efficiency of performing the function internally. Although in recent years outsourcing seems to have taken the lead, astute businesses will pay attention to the cost-reducing effects of artificial invention technology case by case, using artificial invention technology internally and outsourcing externally in those situations in which each is more efficient.

Going one level of granularity deeper, artificial invention technology and outsourcing can be combined with each other in ever more varied ways. For example, "crowdsourcing" refers to a company's use of everyday people using their spare time to create content, solve problems, and even do corporate research and development. Open source software development, in which the goals of a software development project—such as developing a new computer operating system—are made known, and any programmer who is interested can take part, is probably the oldest and most well-known form of crowdsourcing. The Linux operating system has grown to a worldwide phenomenon in this way, with volunteer programmers from every corner of the globe contributing new code and fixing bugs in existing code over the Internet and in their spare time.[13]

What's different about the latest forays into crowdsourcing is that they're being spearheaded by existing for-profit companies seeking to get the best of both worlds: ownership and control over products they develop using the free or low-cost labor of their customers. For example, Eli Lilly created a company named InnoCentive in 2001 "as a way to connect with brainpower outside the company—people who could help develop drugs and speed them to market." InnoCentive, however, does more than simply allow Eli Lilly to tap inventive energy around the world; it offers an "open innovation marketplace" to *any* company willing to pay whoever can solve their problems for them:

> Companies like Boeing, DuPont, and Procter & Gamble now post their most ornery scientific problems on InnoCentive's Web site; anyone on InnoCentive's network can take a shot at cracking them. The companies—or seekers, in InnoCentive parlance—pay solvers anywhere from $10,000 to $100,000 per solution. . . . The solvers are not who you might expect. Many are hobbyists working from their proverbial garage.[14]

These business models differ radically from the traditional corporate aversion to any new technology not developed within the four walls of the company itself—the "not invented here" (NIH) syndrome that we saw earlier. Henry Chesbrough of the University of California at Berkeley has extensively documented the success of companies (including industry giants such as IBM and Intel) that leave the NIH syndrome behind and embrace open innovation by drawing on technological advances wherever they may be found.[15]

There is no need, however, to choose between completely closed and open business models. Combinations of the two models will work best depending on circumstances, and artificial invention technology will enable companies to further develop innovative combinations of open and closed models. For example, we have seen already that existing artificial invention technology cannot make the kind of aesthetic judgments a human can, such as picking the sweeter-smelling of two perfumes. In this case, "interactive evolutionary computation" can be used to combine the best of computing power with the best of human aesthetics. This kind of cooperative technology is a perfect candidate for combining with open business models. Affinnova is doing just that with its Interactive Design by Evolutionary Algorithms (IDEA) software. Cadbury-Schweppes, for example, went to Affinnova when it knew that consumer input was going to be a critical component of the upcoming package design for its 7-Up Plus product. Cadbury and Affinnova broke down elements of the package design into specific components, such as images, colors, package materials, and text. There were literally millions of possible combinations of such components, far too many for any single human package designer to evaluate. This is where Affinnova's IDEA and the power of the crowd came in:

> These . . . millions of package designs . . . were then presented to consumers via an online exercise. During this process, consumers selected their preferred elements to "evolve" the designs. This process leverages Affinnova's technology to mimic the genetic process of "survival of the fittest" and present consumers with more preferred designs, based on previous choices.
>
> At the end of the exercise, Cadbury netted six leading designs from among the millions of potential combinations consumers viewed using Affinnova's evolutionary product design technology. . . . Having a greater understanding of the package elements that appealed to their various audiences, Cadbury was able to choose a final design integrating components from all six "finalist" designs that would appeal to each segment. A solid product paired with a strong package has made 7-Up Plus a strong performer over the last two years.[16]

You probably won't be surprised by now to hear that Affinnova has received at least one patent on its techniques for enabling marketers to use evolutionary computation software to engage customers in the product development process.[17]

Introducing artificial invention technology into the loop accentuates the existing benefits of open innovation by enabling the feedback loop between internal and external actors in the process to function more smoothly and accurately. For example, we've already seen that professional designers tend to be biased about what makes one design better than another. If we relied on those designers to read customer surveys and incorporate them into the company's next design, the designers' biases might still overpower what the customers have said, or blind the designers to a trend that emerges in customer feedback. An artificial neural network or genetic algorithm has no such bias and might actually be able to detect broad-based but indirect trends human reviewers would miss not only because of bias but from an inability to process input from thousands of customers simultaneously.

Don't Wait

As in the case of any paradigm shift, some existing high-tech companies will be tempted to wait on the sidelines and milk their existing inventive processes for as long as possible. They might see artificial invention technology as a fad that will soon pass or that will not bear the cost-reducing or quality-improving fruit it promises. Such companies will be taking a significant gamble in the face of the successes the technology has produced so far. But at least companies already engaging in research and development generally will be aware of the problem and be relatively well poised to retool their processes to use artificial invention technology, even if later rather than sooner.

At higher risk will be those companies currently doing nothing that falls under the rubric of research and development, employing no inventors, and therefore relying solely on external firms to provide any new products and services they may need. Take an industry that is close to my heart (and pocketbook): law. Most law firms purchase all the technology they use, from computers to photocopiers to file cabinets, from external firms. Although they may customize the software applications they use, such customization typically is performed by consultants, and even when performed in-house it falls closer to configuring an existing system than designing a new one. The primary "new product development" that such firms provide is the development of new legal arguments for clients, which is performed by the firms' lawyers using their (highly trained) brains.

Such firms will not be well poised to handle the use of artificial invention software by competitors. Lest you think, as most lawyers do, that software will never be able to develop new legal arguments and advise clients, just read

Richard Susskind's excellent 1998 book *The Future of Law*, in which he chronicles the use of software to answer questions about tax law, draft corporate documents, and offer advice on the convoluted English Latent Damage Act of 1986.[18] If the next big advance in software that provides legal services comes from Lexis-Nexis or another non-law firm provider that chooses to make such software available for use by any law firm willing and able to pay the right price, then at least there is a possibility that the playing field will remain level among firms (or become no more uneven than it is now).

But if one law firm takes the risk of developing such software in-house and succeeds, it could develop a significant advantage over its competitors, whether because the software enables the firm to provide higher-quality legal advice, or offer existing-quality legal advice more efficiently and therefore more profitably, or some combination of both. This just might be a good outcome for the "consumers" of legal services. But it could be a disaster for those individual law firms that have not developed any internal capability at software development; they will be left to scramble to do so in an environment in which the software they need is not available for purchase on the open market. This is particularly true if the innovating law firm patents its software or legal techniques. If you think the practice of law is beyond the reach of patent law itself, just consider that there are already patents on a method of conducting a mock jury trial,[19] a technique for managing outside lawyers,[20] and software for drafting patent applications.[21] (I don't know whether the last of these patents was written using the software that it covers.)

This hypothetical but plausible example of old-fashioned law firms being caught off-guard by a firm that enters the business of inventing illustrates the risks existing companies that are solely consumers but not producers of software are taking, because if a new competitor enters the market who is able to use artificial invention technology to compete, the existing companies will be at a loss to respond quickly. Market forces may be sufficient to respond to this problem, as they appear to have done in response to the growth of the World Wide Web. The advent of the Web as a medium for advertising caught many law firms off guard in the mid- to late 1990s. Law firms have always been laggards in developing new technology, and they were some of the last companies to create feature-rich Websites to make themselves known to the outside world. But they did so eventually, usually by hiring external Web design and hosting firms, without the need to develop any particular internal expertise at Website development. Law firms and other service firms may respond in the same way to the introduction of artificial invention technology in their fields, which would be a boon to artificial invention consultants. But the success of any particular firm depends both on how quickly it mounts a response and how quickly the market makes solutions available. Although law firms seem to have emerged successfully from the economic transformation wrought by the Web, the encyclopedia and newspaper industries are both still reeling and will most likely emerge either radically transformed or not at all.

For better or worse, high-tech companies will find it increasingly difficult to stand by the sidelines and ignore artificial invention, for the same reason companies today in every sector are finding it more difficult to avoid outsourcing their labor. Any company that attempts to continue innovating using old-fashioned human ingenuity alone will face being driven out of the market by competitors using artificial invention technology to invent more effectively, quickly, and inexpensively.

Invention Flat-Earthers Will Suffer

The artificial invention pioneers with whom I have spoken repeatedly encounter resistance to artificial invention technology not because of any technical deficiencies it may have but because of the mere belief that inventing is an inherently human skill, and as a result no computer today or at any time in the future will be able to outinvent a human.[22] Even faced with inventions successfully produced by artificial invention technology, some corporate decision makers are hesitant to accept the facts. Just as some companies have developed a culture that rejects adopting any technology that was "not invented here," so too do many people hold a bias against any technology "not invented by a human."

Bias is the right word for this belief. It is a holdover from a time on the verge of disappearing, much like the way in which previous generations held on to the belief that the Earth was at the center of the universe and humans were the pinnacle of life.

Just as with these other now-discarded biases, those who hold on to this newer version of NIH syndrome will fall behind others who embrace artificial invention technology and the inventions that are produced using it. Perhaps one way to ease the pain of accepting the fact that, at least in some cases, computers can outinvent humans is that doing so does not require us to accept that computers can invent in the same *way* as humans or that computers are conscious or creative, or have any other human trait. The only fact we need accept, and all that will matter for the corporate bottom line, is that computers can produce the *results* businesses need—newer and better products, produced more quickly and inexpensively than before—regardless of how they do it. So even if the NIH crowd turns out to be right about the fact that computers cannot compete with humans when it comes to creativity, they will still find themselves losing out in market competition with those who just don't care and use whatever inventive process produces results most efficiently.

Even legitimate critiques of today's artificial invention technology often exaggerate the technology's limitations. For example, some argue that evolutionary computation technology is not practical for tackling complex problems because evolution, as we know from our real-world experience with life on Earth, takes too long to produce anything useful, and when it does the results are riddled with bugs and inefficiencies. One flaw in such criticism is that it ignores the inefficiencies and failure rates of the design-based techniques against which evolutionary computation must be compared. Furthermore, although evolution can be inefficient at tackling some kinds of problems, it "is actually one of the most efficient ways we know of finding good solutions for many different problems."[23] Finally, artificial forms of evolution are increasingly proving themselves able to overcome the limitations of natural evolution precisely because of their artificiality, which makes it possible for humans to tweak them to achieve efficiencies that natural evolution lacks. For example, although Lamarckian evolution—in which traits that an individual acquires during its lifetime, such as stronger muscles developed by weight-lifting, are passed along to that individual's children—does not occur in the natural world, there is nothing to stop computer scientists from rigging artificial evolution to use Lamarckian evolution so that beneficial traits acquired by one design during its "lifetime" can be passed along to other designs (even to designs other than its children), thereby directing the evolutionary process toward a solution more directly than would be possible using Darwinian evolution alone.[24] Companies that reject artificial invention technology out of hand, on the basis of preconceived notions about evolution or previous generations of the technology, will only suffer from such biases.

13 Advising Aladdin

IF INTELLECTUAL PROPERTY LAW stands to change as much as we have seen, then patent lawyers will have their work cut out for them.[1] The easy part will be for patent lawyers to understand the impact artificial invention technology will have on patent law itself, and to convey this to their clients. The more difficult task will be for patent lawyers to devise concrete legal strategies that will best serve their clients, whether those clients develop artificial invention technology, use it, or merely want to be prepared to defend against suits brought by those who own artificial invention patents. As we shall now see, some of these strategies are far from obvious and stand in stark contrast to those commonly used to protect intellectual property today. As a result, those patent lawyers who merely apply today's approaches to tomorrow's technology will leave clients high and dry.

Writing Patents for Wishes

If artificial genies, wishes, and artificial inventions are all patentable, then patent lawyers will need to learn how to write patents for developments in all three categories. Although patent lawyers will be able to apply many of their existing skills to these new kinds of inventions, such inventions present traps for the unwary.

Consider even the simplest of the three categories: artificial inventions, such as the Oral-B CrossAction toothbrush, the NASA antenna, or John Koza's controller. A patent lawyer could write a patent application for such inventions exactly as if the invention had been invented the old-fashioned way by a human inventor. The patent application would show drawings of the invention and describe how to make and use it.

Difficulties might arise in attempting to describe how to make and use an artificial invention. Once the Creativity Machine devised the crossed-bristle design for the CrossAction toothbrush, for example, one could describe how to use traditional plastic injection molding and other conventional manufacturing techniques to manufacture a toothbrush having the design. This would enable anyone to manufacture and use such toothbrushes and therefore would qualify as a legally sufficient description.

One could, however, describe how to make the toothbrush in another way—by describing how to use the Creativity Machine to produce the design for the toothbrush in the way Stephen Thaler did, by providing existing toothbrush designs and toothbrush performance data to the Creativity Machine—and then describe how to physically manufacture the toothbrush according to this design. One benefit of this approach is that the Creativity Machine might be able to produce designs for a variety of toothbrushes on the basis of the description (wish) laid out in the patent, thereby making the patent broader and hence a more powerful tool for its owner. On the other hand, this approach would require describing how the Creativity Machine itself works (or first making it publicly available in some other way), thereby enabling competitors to use the Creativity Machine freely to produce their own inventions (assuming the Creativity Machine itself was not patented, or that the patent on it had expired). Deciding which approach to pursue requires careful evaluation of the tradeoffs involved.

And that's the easy case. Harder to handle are patents on wishes. Even if you're not a lawyer, you can see how writing a patent for an equation that governs airflow around an automobile frame is a different task from writing a patent for an automobile frame itself. An airflow equation patent necessarily describes the equation itself, how it characterizes real-world airflow, and how a description of the equation can be written in a language an artificial genie can process as a wish to produce specific designs for new and useful automobile frames. A patent lawyer who writes such a patent application therefore needs many of the same skills as artificial inventors themselves, namely a grasp of abstract mathematics and physics and an understanding of the wishing languages used by artificial genies.

A traditional patent on an automobile frame, in contrast, would show drawings of the automobile frame itself from various angles and describe the shape of the frame, the materials out of which it is constructed, and the best way to manufacture the frame. Although the patent could also describe the airflow equation, it might not. The best patent attorney to write such a patent application is someone with a background in mechanical engineering and technical specification writing.

Once again, we need only look to the shift that has already been caused by "traditional" software to see such a modification in the skills needed by patent lawyers. Take a look at a claim for a lever, slightly modified from the treatise *Landis on Mechanics of Patent Claim Drafting*:[2]

A lever comprising:
    a forked end, the forked end comprising spaced apart
furcations; and
    a pivot pin mounted between the furcations.

(In case you were wondering, a "furcation" is the place where something divides into branches.)

This is a classic patent claim for a mechanical device. It lists the *structural elements* of the lever and describes how they are connected to each other in a "knee bone's connected to the thigh bone" fashion.

Here, in contrast, is a claim from the patent that I mentioned earlier on a method for writing a patent application:[3]

A method by computer for drafting a patent application having at least sections including claims, a summary of the invention, an abstract of the disclosure, and a detailed description of a preferred embodiment of the invention, said method comprising the steps of:

requesting and storing primary elements (PE) of the invention that define the invention apart from prior technology before drafting the claims;

drafting the claims before drafting the summary of the invention, abstract, and the detailed description of a preferred embodiment of the invention; and

drafting the sections in a predetermined order prohibiting jumping ahead to draft a latter section.

This claim, like the claim for the lever, lists several elements. These elements, however, are *steps* (subtasks) performed in a method, rather than physical components of a device. You can think of the method described by this claim as a recipe that includes a sequence of steps that the software performs, namely, "requesting and storing primary elements," followed by "drafting the claims" and "drafting the sections." This mirrors the way in which the majority of software code itself is typically written. This format is typical of method claims in general and software patent method claims in particular.

Although patent law allowed patents to be granted for processes long before software existed, historically most process patents were for chemical processes, and software patent claims are the first that typically lack any explicit reference to physical actions. A patent claim on a physical process, such as one for making flour, will describe physical actions such as harvesting the wheat, separating the wheat from the chaff, milling the wheat, and so on. These are clearly physical acts performed on physical components. A software patent claim, however, such as the one above, often directly refers only to actions performed on numbers, data, records, fields, and other objects that could be *either* physical *or* abstract entities. Although when a computer performs such operations it does in fact perform physical actions by manipulating electrical signals within its processor and memory, such physical actions are only *implicit* in the language of many software patent claims.

These features of software patent claims make the process of *writing* such claims, and the skills needed to do so, as different from writing claims for mechanical devices as computer programming is from mechanical engineering. The two processes have something in common, to be sure, and software patent lawyers need to learn the basics of patent claim drafting drawn from centuries of patent law before adapting those skills to writing software patent claims. But those whose only background is in mechanical or even electrical engineering and in patent drafting as it applies to those fields often find it difficult to write software patents because of the more abstract way in which software patents need to be written.

Patent lawyers who want to help their clients obtain patents on wishes would do well to learn from this history of transition from the traditional structural type of claim drafting, which characterized patents on mechanical devices for most of patent law's history, to the procedural type of claiming that characterizes patents on software. The format of patent claims, in other words, has roughly tracked the manner in which inventors in their respective fields think about and describe their own inventions. If history is any guide, therefore, patent lawyers should start to learn to think about and write patent applications in the wishing languages of artificial invention technology.

There is an opportunity here for patent lawyers to be in the forefront of educating their clients that it is even possible to obtain patents for wishes. Even now, long after patent law has embraced traditional software, many potential clients ask me whether it is possible to patent software or whether they must rely on copyright and trade secret law for protection. Although software patents have been granted in large numbers in the United States for almost 20 years (and in smaller numbers for much longer than that), the word can be slow to get out to those who stand to benefit from it. The astute patent lawyer will be sure to let clients know that the future is already here.

Patent lawyers who are already skilled at obtaining patents on software will be well poised to help their clients obtain patents on wishes. After all, even traditional software is a kind of wish, in the sense that it describes a procedure for a computer to perform. As we have already seen, artificial invention technology enables such wishes to be written at even higher levels of abstraction.

But this doesn't mean existing strategies for obtaining software patents can simply be adopted without modification as strategies for obtaining patents on wishes. To offer just one example of how patent lawyers will need to be cautious, in general it is sufficient for a patent lawyer writing a software patent application to write a functional description of the software, that is, a description of what the software *does*.[4] This is because with most traditional computer programming languages, once a program is written describing what the program does, it is guaranteed that the program will run and perform the function it describes, so long as the programmer followed the rules of the programming language. The ability to transform a written computer source code into working software is predictable in this sense.

Not so with much of the latest batch of artificial invention technology. Give a wish to a genetic algorithm in the form of a fitness function for car frame airflow, and the genetic algorithm may or may not produce a useful car frame. It may just grind to a halt without producing anything that satisfies the airflow equation (wish) to anyone's satisfaction. Therefore, merely offering a description of a wish in a patent application may not be legally sufficient to demonstrate that one has invented anything useful. Given the uncertainty about which legal rules will develop to govern such cases, it will probably be safest for the practicing patent lawyer to advise his clients to test their wishes and furnish proof they can produce useful results so that such proof can be included in the patent application itself.[5] This would make patents on wishes resemble a mix of existing software patents, which are written abstractly, and existing chemical and biotechnology patents, in which unpredictability requires proof that a useful result has actually been obtained. Attempts by patent lawyers to blindly apply existing legal strategies to wish patents will be fraught with danger.

The final case—patents on improvements to artificial invention technology itself—are difficult not because it will be particularly hard to write patents for such improvements, but rather because it is not easy to decide whether to pursue patent protection for such technology at all.

Reevaluating Patents vs. Trade Secrets

The big-picture concept that patent lawyers will need to grasp and convey to their clients is that although today patent law determines who owns *inventions*, tomorrow patents on artificial invention software will determine who owns *the ability to invent*. Patent law, therefore, will confer even more awesome power on those who take advantage of it in the Artificial Invention Age than it does today.

What I mean by owning the ability to invent is that until now the ability to invent was lodged in the minds and bodies of human inventors. To a certain extent, this kept the ability to invent free from the ownership of anyone except inventors themselves, at least in legal systems that disallow slavery. Inventors were generally free to use their inventive talents however and for whomever they chose. To the extent inventors sought to market their skills for the highest dollar, market forces and government funding were the primary influences on where inventors devoted their energies.

The law allowed some degree of ownership over the ability to invent, however. Contract law and employment law sometimes allow inventors and their employers to enter into noncompetition agreements, whereby an inventor who works for a particular company agrees not to work for any of that company's competitors for a certain period of time, say, one year. Such agreements give the inventor's current employer some degree of ownership over the inventor's ability to invent. The law generally imposes limitations on such agreements, however, out of a general interest in keeping inventors free to work for whomever they choose.[6]

Trade secret law also allows a company to retain a certain kind of owner-ship over the company's accumulated internal inventive skill, referred to as know-how. Although you might think of trade secret law as attaching to se-cret formulas (such as the one for Coca-Cola), a company can use trade secret law to protect any information that gives the company a competitive advan-tage and that is not generally known to the public.[7] Inventors in any company who spend years inventing new products for the company will not only invent those specific products but also generate a certain body of knowledge and techniques for inventing. This knowledge might be stored in the form of for-mal written policies, informal best practices passed from inventor to inventor through personal interactions, and even attitudes and confidence inspired by leaders within the company. This "secret sauce" of the company—its collec-tive inventive skill—may be its most valuable asset and is legally protectable, if at all, by trade secret law.

Although certain processes a company uses to invent more effectively than its competitors might technically be patentable, any company that relies on inventive prowess for its competitive advantage should think long and hard before attempting to patent those processes because obtaining such a patent would require the company to reveal to the world how they work. Recall that patents are published for the world to see. Although a company might de-cide in a particular case that it would benefit more from patent protection than trade secret protection on a particular method of inventing, the bean-spilling effect of patent protection should at least cause the company to care-fully weigh trade secret against patent protection before deciding to apply for a patent. Historically at least, the norm has been for companies to keep their accumulated inventive knowledge under wraps as trade secrets and obtain patent protection on the products and processes they sell to the public. It's like keeping the goose that lays golden eggs as a trade secret and patenting the eggs. The result is that the company gets the best of both worlds: because trade secrets never expire,[8] the company can use its accumulated inventive skill to churn out inventions *indefinitely* while using patent law to protect the inventions themselves against copying by competitors.

This is the first step down a path that may lead the unwary patent lawyer or high-tech company into a pitfall. Although trade secret and copyright were the dominant means of legal protection for software through the 1980s, com-panies have increasingly turned to patent law to protect their software for the last two decades.[9] Some companies, and the patent lawyers who advise them, seem to have adopted what appears to be a knee-jerk "patent everything" approach to any new software they develop that satisfies the legal requirements for patentability.

Such a strategy could result in disaster for companies pursuing it for artificial invention technology they develop. Artificial invention technology is a bottled genie, an inventive process embodied in a tangible programmed computer. This is what makes it patentable. But if it is true that trade secret generally is the most profitable way for companies to protect their inventive skill, and if artificial invention software is "inventive skill in a bottle," then by default any inventor or company that develops new artificial invention software should keep it as a trade secret rather than patent it. This way they can use the technology to churn out new inventions exclusively and sell them publicly *forever* and without letting anyone else know how the technology works, because trade secrets do not expire as patents do.

Let's consider each of these advantages of trade secret protection in turn. First, a trade secret lasts for as long as its owner takes reasonable steps to keep it secret,[10] as the continued trade secret status of the formula for Coca-Cola attests. A patent, in contrast, expires a mere 20 years after the application for the patent is filed.[11]

Second, because in most cases selling an artificial invention, such as a toothbrush, gives its buyer no information about the artificial invention technology that was used to invent the toothbrush, patenting the toothbrush and keeping the genie secret is a particularly good strategy. By way of contrast, consider that one reason patent protection has been such an effective way to protect much traditional software is that making the software available to the public often renders it trivial for anyone with programming skill and access to the software to understand how it works and copy its functionality. For this reason, software has been said to carry "its know-how on its face."[12] As a result, anyone who sells software runs the risk that the customer will be able to reverse-engineer the software, thereby defeating trade secret protection, which does not protect against reverse engineering.[13] Patent law, on the other hand, does allow the patent owner to prevail even against someone who reproduces the functionality of the patented software by reverse engineering it.[14] As a result, patent protection is often a better way to protect traditional software than trade secret protection.

But these considerations don't apply to the toothbrush, because its purchaser can inspect the toothbrush for as long as he pleases and still gain no insight into how to produce his own genie for inventing a toothbrush (assuming that the original genie has been kept secret). Therefore there is no special need to use patent law to protect *secret* toothbrush-inventing genies against reverse engineering. This is yet another example of how artificial invention technology may be used in a way that is analogous to companies' traditional internal use of human inventors. In both cases trade secret law is preferable to patent law for protecting the inventive process. These reasons indicate that developers of artificial invention technology will have strong incentives to maintain their refinements to such technology as trade secrets.

Furthermore, if you develop a new kind of genie and patent it, you stand to reap all the benefits of the patent, but at the same time you will raise the level of ordinary skill in the art and therefore potentially make it more difficult for you to prove that your own subsequent artificial inventions are nonobvious and therefore patentable, for all the reasons we saw in Chapter 7. In other words, patenting or otherwise publicly disclosing a new genie may make it harder for everyone—including you—to obtain patents on any wishes such a genie is capable of granting. If, on the other hand, you keep the genie a trade secret, you will make it easier for yourself to obtain patents on any wishes and artificial inventions you subsequently create.

This is not to say it will never make sense for someone to patent a new kind of artificial invention technology. Patents confer some significant benefits over trade secrets. You might benefit from patenting your new genie if you want the ability to stop your competitors from using it, even if they independently invent the same genie themselves (that is, without copying yours), because patents protect against such "independent invention" while trade secret protection does not.[15] My point is not that artificial invention technology should never be patented, but rather that its value as a *tool for inventing* and not merely as an end product itself makes it particularly important to evaluate the tradeoffs between patent and trade secret protection very carefully in individual cases, instead of automatically seeking patent protection for every new breed of artificial invention technology out of a misguided notion that patents are always the best way to protect software.

Any good patent lawyer should already be working closely with his or her clients to perform evaluations of this kind for new technological developments.[16] The consequences of getting the answer wrong for artificial invention technology, however, are potentially more severe than for most other kinds of software. Patent lawyers therefore need to be even more attentive to the nature of the technology developed by their clients and also to the business goals those clients seek to achieve by developing the technology.

Patent lawyers who have advised clients who develop other kinds of tools of invention should already have learned this lesson. Artificial invention technology is not the first kind of invention augmentation tool; just think of the microscope, the test tube, and the computer itself. Although many versions of these technologies obviously have been patented and therefore made public, companies continue to make custom modifications to these and other tools and then maintain those modifications as trade secrets for their own private use in inventing. Companies, with the assistance of their lawyers, decide whether to seek patent or trade secret protection in a particular case depending on the company's business goals. As a simple example, a company that specializes in selling microscopes to laboratories will probably seek to patent the next microscope it develops, while a laboratory that profits from inventing and selling new chemicals might choose to keep the new microscopes it designs as trade secrets. Two different companies, in other words, might choose contrasting forms of legal protection for the same invention according to their business models.

Although at a fundamental level the choice between patent and trade secret protection for artificial invention technology may be based on the same considerations as for these older forms of invention facilitation tools, the difference is that artificial invention technology comes significantly closer to embodying the entire process of invention in a closed-loop system than does a microscope. As a result, a company that sells a microscope doesn't necessarily enable its competitors to invent the next generation of microscope more easily, while a company that makes public its artificial invention technology gives competitors the very tool they need to invent its successor.

The Artificial Invention Age will challenge patent lawyers to return to first principles and reevaluate their time-tested legal strategies in light of artificial invention technology. In addition to learning about how artificial invention technology works, patent lawyers must learn how to write patent applications that cover artificial genies, wishes, and artificial inventions both accurately and broadly. The Artificial Invention Age will challenge patent lawyers to the limits of their ability, but it will also keep them very busy, and for as long as lawyers bill their clients by the hour, patent lawyers will welcome the Artificial Invention Age with open arms.

14 A Genie in Every Home

USERS OF COMPUTER TECHNOLOGY in the Artificial Invention Age will benefit from better, cheaper products brought to market more quickly than ever before. Most consumers will be satisfied to just sit back and enjoy the ride. But some "mere" consumers will see an opportunity to leverage artificial invention technology and become inventors themselves. Artificial invention technology stands poised to bring invention to the masses—to democratize invention.[1]

Inventing on the Cheap

We've already seen how artificial invention technology can radically reduce the cost of inventing. Technology that reduces the cost of producing and distributing new creative works can, however, do more than boost profits for the producers and lower the cost of products for consumers; it can also expand the class of people who have the skills needed to create new products in the first place. By lowering the cost of publishing, the Internet brought authorship to the masses (think blogs and Wikipedia). Low-cost recording technology and peer-to-peer networks brought music publishing to the masses. Low-cost computer hardware and high-level programming languages effectively brought electrical engineering to the masses.

Although you may balk at that last example, we already saw in Part I how computer programming is a kind of shortcut for rewiring a computer. In this sense, low-cost personal computers have enabled a larger number and broader range of people who lack any training in electrical engineering to create software capable of performing functions that, in the absence of computers, would have required electrical engineering to achieve. Although conceiving of programming as a kind of electrical engineering may require us to have a broader understanding of that which constitutes "electrical engineering," this is similar to conceiving of the volunteer authors of Wikipedia as encyclopedia publishers even though none of them is traveling door-to-door, shipping physical books, or engaging in many other activities required to satisfy the traditional definition of "encyclopedia publishing."[2]

I happened to stumble onto personal computers shortly after the first versions of such computers that were affordable for the average consumer entered the market in the late 1970s. The first computer our household purchased, an Atari 800, sold for about $500. Just a few years earlier, minicomputers from companies such as DEC sold for $10,000 and up, themselves a significant drop from the million-dollar computers of the previous generation. Little did I know at the time that the plummeting cost of computers, combined with the increasing power of high-level programming languages and the low cost of copying and distributing software, had already made it possible for individuals or pairs of programmers to write the programs that formed the foundation of some of the leading software companies of the day. For example, the first spreadsheet software for personal computers, VisiCalc, which "is often described as the 'killer app' that unleashed the corporate personal computer revolution,"[3] was written by just two people: Dan Bricklin and Bob Frankston.[4] Similarly, two of the earliest word processors for personal computers, Electric Pencil and EasyWriter, were written by lone authors in 1976 and 1977.[5] Ken Thompson wrote the first incarnation of the Unix operating system from his house in about four weeks in the summer of 1969.[6] Many of the earliest industry-specific personal computer software applications were written by "physicians, dentists, or insurance brokers"—not trained as engineers or programmers—for use in their own industries.[7]

Such accomplishments by individuals, often lacking technical training, has not been possible in fields such as pharmaceuticals and nuclear power, where the high costs of entry and the lack of artificial genies prohibit any individual outside of a major corporation or university from gaining access to and obtaining firsthand experience with the necessary technology. This is poised to change as the cost of computing power continues to drop exponentially.[8] A computer with the power of the supercomputer that John Koza built in 1999 for $450,000[9] could be purchased today for about $7,000. If Moore's Law—which essentially states that the computing power you can buy for one dollar doubles every 18 months—continues to hold for another ten years, then it will be possible to buy a computer with the same processing power for under $100 in 2019. Put another way, in 2019 Koza's $450,000 will buy a computer that is about 8,000 times more powerful than the one he built in 1999. Such developments have the potential to bring inventing to the masses in fields ranging from robotics to materials science to pharmaceuticals in the same way that previous generations of affordable computer technology made it possible for individuals to create commercial-quality software.

We can already see the seeds of this democratizing effect within the world of technologists. Dr. Koza has been using artificial invention software to invent lenses,[10] even though he's acknowledged that he knows almost nothing about optics.[11] Although he is a gifted scientist and computer programmer, artificial invention technology effectively augments his skill to enable him to invent in fields where he has no formal training. Although this is not yet inventing for the masses, it represents a significant step in that direction by expanding the range of fields in which inventors are capable of inventing.

Synergy with Open Innovation Models and the Rise of the Prosumer

We already saw in Chapter 12 that businesses are using open innovation to reach beyond the four walls of the corporation to improve their products and services. Although in some cases this means turning to consultants and other companies, increasingly it means looking to the company's own customers not just for traditional demographic information but for real inventive input. Consumers increasingly are becoming active participants in product design, whether to give back to the manufacturer or simply to improve products for their personal enjoyment. Owners of consumer electronic devices from the Microsoft Xbox to the Sony Playstation, iPod, and Tivo, dissatisfied with the limitations of those products, have cracked open the boxes and put in faster processors and made other unauthorized "mods" to tweak the products as they see fit.[12] This new breed of proactive user, who acts as both producer and consumer, is now known as the "prosumer."[13]

As Eric von Hippel of MIT points out, however, the prosumer is not so new as we often think. He documents how the entire sport of high-performance windsurfing was spurred by an avid windsurfer who added footstraps to existing windsurfing boards in the late 1970s to enable himself to jump and land without losing contact with the board.[14] More generally, von Hippel has found that consumers (including individuals and firms) develop or modify products for their own use with a frequency of 10–40 percent across a variety of product types.[15]

Although some companies initially reacted hostilely to such activity, even taking legal action against those who sold components for making unauthorized modifications to their products,[16] forward-thinking companies have begun to realize they stand to benefit from the hard work and ingenuity of their customers, particularly because such customers are often in a better position to gauge their individual and rapidly evolving needs than the company's own engineers are.[17] As a result, we are already beginning to see companies actively taking prosumers under their wing and incorporating the improvements they make into the next generation of their products.[18] For example, the online movie rental company Netflix has offered a $1 million prize to anyone who can create an algorithm for recommending movies that can outperform Netflix's own "Cinematch" algorithm by at least 10 percent. Contestants submitted more than a thousand entries within the first month alone. Even though no one has won the prize yet, Netflix has already paid one team of contestants for the right to incorporate improvements from their submissions into Cinematch.[19]

Companies are going so far as to sell products that are specifically designed to be modified by customers. For example[20]:

- Illah Nourbakhsh of Carnegie Mellon University has developed a new robotics platform called Qwerk that Nourbakhsh "hopes will launch an open-source robotics movement and 'democratize robot design for people intimidated by current techniques and parts.'" Qwerk includes low-cost components, including the hardware and software necessary to connect to the Internet, which people can interconnect to create their own robots in just a few hours.[21]

- Microsoft's Studio Games Express lets customers develop their own games for the popular Xbox 360 gaming console. Gamers can share their games with each other by posting them to the XNA Creators Club—for a $99 annual fee.[22]

- iRobot released the Create robot, a toolkit that customers can use to create their own robots. iRobot came up with the idea after discovering that customers were hacking its robot vacuum cleaner to create their own robots. iRobot maintains a message board where customers can discuss their latest creations, such as a robot for retrieving beer from the refrigerator.[23]

The growing ability of artificial invention technology to enable consumers to become inventors will only amplify these existing trends, providing further incentives to businesses to work with, rather than against, their customers in innovating. This will be yet another reason for existing in-house inventors to begin learning how to leverage artificial invention technology now, because they will increasingly compete not only with inventors in their own and other firms but also with a growing mass of consumer-inventors working as professional freelancers.[24] This is exactly what happened in the early years of the software industry, when costs of entry were so low that "all you need[ed was] two programmers and a coffee pot" to start a software company if you had the necessary programming skills.[25]

As we have seen, companies currently turn to consumers most frequently in cases where the customers' human expertise and aesthetic judgment are needed. But the democratization of invention that artificial invention technology makes possible could make consumers valuable in other situations as well, as full-fledged coequal inventors. Although I don't know of any examples of it just yet, there is nothing to stop customers themselves from using artificial invention technology, rather than their personal judgment, to make improvements to products and send those improvements back to the company for use in its subsequent offerings.

Synergy with Personal Fabrication

Even if artificial invention technology makes it possible for the Average Joe and Jane to produce blueprints for their inventions, the revolution would not be complete if a large and expensive industrial manufacturing plant were still necessary to build working versions of those inventions. As MIT Professor Neil Gershenfeld points out in his book *Fab*, however, "personal fabrication" technology is already available at low cost; his students have used it to design and build everything from bicycles to chairs to an alarm clock that makes you solve a physical puzzle to turn it off.[26] According to Gershenfeld, a personal fabricator (PF) is

> a machine that makes machines; it's like a printer that can print *things* rather than images. By personal fabrication, I mean not only the creation of three-dimensional structures but also the integration of logic, sensing, actuation, and display—everything that's needed to make a complete functioning system. With a PF, instead of shopping for and ordering a product, you could download or develop its description, supplying the fabricator with designs and raw materials.
>
> Programmable personal fabricators are not just a prediction, they're a reality. The world of tomorrow can be glimpsed in tools available today.[27]

Personal fabrication closes the loop on the automation of invention; use an artificial genie to turn your wish into a design for a concrete product, and then use a personal fabricator to build the product itself. Personal fabrication for the masses is the last link in the chain needed to connect wish to design to physical reality, for complete end-to-end democratized inventing.

Computers always have been able to do this, albeit for a smaller range of wishes. As you now know, when you write a high-level program and program a computer with it, the computer both translates your human-readable instructions into a detailed software design and modifies its own memory to implement the design. The fact that computers perform *both* of these steps in an integrated process, and in a way that is internal to the computer and not directly visible to the user, is why I think the genie metaphor has eluded us for so long. John von Neumann, however, recognized the possibility that a computer, as a universal machine, could be connected to a "universal constructor" for controlling matter *outside* the computer.[28] Although the technology of his time did not allow him to build such a constructor, he would surely be smiling today if he could see where artificial genies and personal fabricators have taken us.

What's Old Is New

The combination of artificial invention technology, personal fabrication, and open innovation models will bring new power to consumers not only to demand but also to design, build, and share their own inventions. The technological and economic forces that led businesses to favor mass manufacturing of one-size-fits-all products at the tail end of the Industrial Age are disappearing.[29] Even the currently touted trend of mass customization, according to which products such as computers, windows, and blue jeans can be tweaked on a per-customer basis,[30] will pale in flexibility in comparison to technologies and business models that truly put the consumer in the driver's seat. Individuals will once again have not only the tools but also the economic incentives to exercise their full range of creative abilities, rather than merely the important but narrow one of choosing among a list of cookie-cutter alternatives designed with only their coarsest preferences in mind. As Eric Bonabeau of Icosystem puts it:

> I think this grand synthesis of evolution and design will become more and more pervasive in engineering, in marketing, in consumer product design. I think that what we will see in ten years is people designing their own stuff. They will have the power to do that, because they will be able to evolve whatever they want without knowing anything about the technology underlying the design process. They don't need to know anything about designing a car; all they need to know is what kind of car they want.[31]

As Gershenfeld points out, "The invention of industrial automation meant that a single machine could now make many things, but it also meant that a single worker who used to do many things now did only one."[32] Artificial invention technology and personal fabrication stand poised to correct this historical error by enabling a return to the renaissance artisan,[33] who can see a problem through from initial conception to final construction of a working machine by combining a broad knowledge of science and arts with boundless curiosity and razor-sharp problem-solving acumen. "Such a future really represents a return to our industrial roots, before art was separated from artisans, when production was done for individuals rather than masses."[34] A return to the days of Leonardo da Vinci, except that this time Leonardo will have a genie at his right hand.

Acknowledgments

NO COMPUTER AUTOMATED the writing of this book. To the contrary, the work before you is the result of a seven-year effort completed with the assistance of countless family members, friends, colleagues, and others who offered not only constructive feedback but critical insights, direction, and encouragement at every step along the way.

First and foremost, my partner, Melissa Hoffer, began this journey with me at our breakfast table back in the summer of 2001, when I read an article about software patents (as I often do at the breakfast table) and bemoaned the inconsistent usage of terms such as *software* and *computer program*, which led me to believe that much of the confusion about software patents could be dispelled simply by clearly answering the question, "What is software?" Since that day, Melissa has engaged me tirelessly and in countless discussions about invention automation and patent law whenever I have felt compelled to raise the topic, no matter how inappropriate the circumstances—at dinner, on road trips and airplane flights, while grocery shopping and weeding the garden, during quiet walks through the woods, and while gazing at stars. Her input, feedback, patience, and support have been so consistent a feature of my writing experience that I could not even attempt to isolate their impact to any particular portion of the final text.

Many research assistants made critical contributions not only to the book's main text but also to its most underappreciated feature: the endnotes. Meleena Bowers's careful and comprehensive research laid the foundation for my first papers on software patents, which informs much of Part II of this book. Cynthia Gilbert went above and beyond the call of duty by clearly summarizing for me such diverse topics as First Amendment protection for computer source code and the ethics of distributing harmful software, as well as contributing insights from her former career as a computer security professional. Brandy Karl, now a Fellow at Stanford Law School's Center for Internet and Society, was already more of a colleague than a research assistant when she helped me prepare the first draft of the proposal for this book and create my blog on invention automation. Irena Zolotova covered territory ranging from Edison and the light bulb to the software patent lawsuits making today's headlines. Christopher Albert tackled complex topics in patent law—ranging from the fuzzy and fluid contours of nonobviousness to the ongoing legislative proposals for patent reform—with ease. Jonathan Varsanik's research on the technology of invention automation made it possible to transform Part I of this book from a philosophical discourse on computerized genies to an exploration of how real technology is being used to automate inventing today.

My agent, Jodie Rhodes, championed this project with such enthusiasm from the day she received my initial query letter that she single-handedly renewed my faith in the publishing industry. Jodie's incredible persistence in marketing my proposal to publishers and negotiating on my behalf sometimes makes me wonder which of the two of us has put more time into bringing this book to press.

Randy Burgess worked closely with me to transform my original book proposal from an academic treatment of the relationship between computer technology and patent law into a compelling case for the need to publish a book on the effects invention automation technology will have on real people—while bolstering, rather than sacrificing, any of the theoretical insights with which I began. He provided constructive criticism and specific suggestions on drafts of the book itself, which kept me from straying from the proposal's roadmap and also led to further improvements in clarity, readability, and logic at every level, from the organization of the book's major sections down to the wording of individual sentences.

I thank Amanda Moran, my original editor at Stanford University Press, for having faith in a book by an unknown author on an unknown topic. She was succeeded by Kate Wahl, whose feedback on multiple drafts of the book helped me say more with fewer words and clarify complex concepts for a diverse audience. She also showed unending patience in answering all of my first-time-author's questions about the ins and outs of the publishing process.

A small army of volunteers generously reviewed drafts of the book and the academic papers preceding it. In particular, Steve Gold pressed me to clarify my use of terms such as *automation* and *invention*, leading to significant improvements in the discussion in Part I of inventing as a process of interaction between human and machine; any remaining ambiguities are solely my own. Steve's sketch on a legal pad over coffee at a Starbucks one morning led to the critical Figures 9 and 10 in Chapter 6.

Richard Goldhor's careful reading of and detailed feedback on the book's first draft let me know both where I had succeeded and where my arguments were internally inconsistent, not supported by the facts, or unnecessary to support my conclusions. I also thank Rich for introducing me to Amy McCreath at the MIT Technology and Culture Forum, who graciously invited me to speak on the risks and ethical implications of invention automation.

Ralph Clifford, an authority on computer authorship and copyright law in his own right, took the time to supply gentle but pointed feedback on the first draft of the book that helped me avoid making errors that only someone with his experience would notice.

Karen Christensen spent the better part of a day explaining the ins and outs of the book proposal process to me. I doubt I would ever have found an agent, much less a publisher, without her early guidance.

I received invaluable feedback on my early academic papers and conference presentations on invention automation from many others, notably Hal Abelson, Archon Fung, Peter Gordon, Larry Kolodney, and Michael Martin.

I had the opportunity to hone many of the legal and philosophical arguments at the heart of this book through presentations at conferences and other venues sponsored by a variety of organizations over the years, including the Institute of Electrical and Electronics Engineers (IEEE) Society on Social Implications of Technology; the Dartmouth Lawyers Association; the Nelson A. Rockefeller Center at Dartmouth College; the British and Irish Law, Education, and Technology Association (BILETA); the Society on Social Implications of Technology of the Boston Section of the IEEE (special thanks to Alex Brown); the North American Computers and Philosophy Conference; the Works-in-Progress Intellectual Property Colloquium; the Information Ethics Roundtable (special thanks to Don Fallis, Kay Mathiesen, Catherine Womack, and Tony Doyle); the Centrum voor Wiskunde en Informatica (special thanks to Jan Bergstra and Paul Klint); the Boston University School of Law Faculty Workshop (special thanks to Michael Meurer); and the Stanford Center for Computers and the Law (special thanks to Brandy Karl, Harry Surden, and Joshua Walker).

I extend particular thanks to Jim Moor of Dartmouth, who has provided me with feedback and encouragement from the very first conference paper I presented at the International Symposium on Technology and Society in 2002. I often find myself humbled when I formulate a "new" thought about the philosophical implications of computers only to find that Jim already expressed the same thought, with deeper insight and subtlety, a decade or two earlier.

In writing this book, I had the pleasure and the privilege to interview many of the pioneers in the field that I refer to as artificial invention technology. They included (in alphabetical order) Sion Balass, Cem Baydar, Peter Bentley, Hans-Georg Beyer, Eric Bonabeau, Lawrence "David" Davis, Robert Felton, David Fogel, James Foster, David Goldberg, J. Storrs Hall, John Holland, Gregory Hornby, Lorenz Huelsbergen, Martin Keane, Didier Kaymeulen, John Koza, Hod Lipson, Nicholas Macias, David Montana, Jordan Pollack, Joe Rothermich, Karl Sims, Lee Spector, Adrian Stoica, Christof Teuscher, Stephen Thaler, Marcel Thuerk, Andy Tyrell, Jeff Walker, and Ricardo Zebulum. Without exception, they were generous with their time and patient in their tutelage, immediately assuaging my initial concern that they might be hesitant to open up to a patent lawyer knocking on their doors from out of the blue.

My business coach, Van Smick of ActionCOACH, enabled me to keep my patent law practice running smoothly and my clients happy throughout the entire book-writing process. He helped me manage my own time efficiently as well as systematize many—and even automate some—of my firm's internal processes so that I could dedicate more of my time to serving my clients and writing this book, and less time to managing my practice's day-to-day operations.

Thanks to Cathy Spinney of Spinney Associates for the illustrations of computers, wishes, and genies dispersed throughout the book. Cathy transformed my rough sketches, and in some cases just my brief descriptions, into visual realizations of key concepts that are (to my graphically oriented mind) both clearer and more concise than any text could be.

Last, but not least, my dog, Maggie, not only lay faithfully at my feet during countless extended writing sessions but served as an ever-reliable sounding board for new ideas and encouragement to get out of my chair every few hours to go for a mind-clearing walk. If I didn't know otherwise, I would swear she spoke back to me whenever I voiced frustrations about my latest roadblock and encouraged me to push forward.

Excerpted text is reprinted with permission as follows:

Fab: The Coming Revolution on Your Desktop—from Personal Computers to Personal Fabrication, by Neil Gershenfeld, 2005. Reprinted by permission of Basic Books, a member of Perseus Books Group.

From Airline Reservations to Sonic the Hedgehog: A History of the Software Industry, by Martin Campbell-Kelly, Cambridge, MA: MIT Press, 67, 31–35, 98, 90. © MIT 2003. Reprinted by permission.

"Early and Often: Harnessing the Consumer's Voice in Packaging Design," by Robert Wallace, *Package Design,* November/December 2005. Reprinted by permission.

Patent Failure: How Judges, Bureaucrats, and Lawyers Put Innovators at Risk, by James Bessen and Michael J. Meurer, 2008, Princeton University Press. Reprinted by permission of Princeton University Press.

Genetic Programming IV: Routine Human-Competitive Machine Intelligence, by John R. Koza et al., 2003. Used with kind permission of Springer Science and Business Media and John Koza.

Interview with Eric Bonabeau in *Technology Review.* Reprinted by permission of Eric Bonabeau.

Notes

Introduction

1. Similarly, University of Illinois at Urbana-Champaign Engineering Professor David Goldberg argues that "the 'real' information revolution awaits effective computational aid for the heavy lifting of thought. Although computers now keep our books and crunch our numbers, they have not been particularly helpful in assisting our innovation and invention. This is about to change." Professor Goldberg refers to the upcoming age as a "golden age of computational innovation." David E. Goldberg, *The Design of Innovation* (Boston: Kluwer, 2002), 227.

2. I use the term *artificial invention technology* throughout this book to refer to a variety of technologies, such as genetic algorithms, artificial neural networks, hardware description language synthesizers, and even traditional computer programming language compilers, to the extent that those technologies can be used to automate inventing. Although technologies that fall under the heading of artificial invention technology may be used to perform tasks other than inventing, in this book I focus only on their use in inventing.

3. Tina Hesman, "The Machine That Invents," *St. Louis Post-Dispatch*, January 30, 2004.

4. Three model ST5 33.142.7 antennas are in space on the ST5 mission. Gregory Hornby, interview by author, October 12, 2007; Gregory S. Hornby, Jason D. Lohn, and Derek S. Linden, "Computer-Automated Evolution of an X-Band Antenna for NASA's Space Technology 5 Mission," *IEEE Transactions on Evolutionary Computation* (forthcoming).

5. Gregory Hornby, interview by author, June 28, 2005.

6. U.S. Patent no. 6,847,851 (issued January 25, 2005); U.S. Patent no. 7,117,186 (issued October 3, 2006).

7. The design for Dr. Koza's controller was generated automatically (without human intervention) in one pass by software. John Koza et al., *Genetic Programming IV: Routine Human-Competitive Machine Intelligence* (Boston: Kluwer, 2003), 102–104. Not every design, however, was generated completely automatically. For example, the NASA antenna design was generated using multiple passes, with the human engineers learning from initial output of the software and feeding results of such learning back into the software to improve its output. Gregory Hornby, interview by author, June 28, 2005. The advantages and disadvantages of such differing approaches will be described in Part I.

8. Stephen Thaler, email message to author, February 17, 2008, and interview by author, February 20, 2008.

9. According to University of Michigan Electrical Engineering and Computer Science Professor John Holland, the role of scientists and engineers who use evolutionary computation today is to "pose the question," rather than furnish the answer. John Holland, interview by author, September 27, 2005.

10. The actual problem description (in this case, a "fitness function") that was used to evolve the NASA antenna shown in Figure 1 may be found in Jason D. Lohn et al., "An Evolved Antenna for Deployment on NASA's Space Technology 5 Mission," Genetic Programming Theory Practice 2004 Workshop (GPTP-2004), May 2004.

11. K. Eric Drexler used the term *genie machine* to refer to AI systems that "have both great technical ability and the social ability needed to understand human speech and wishes": "What you ask for, it will produce." K. Eric Drexler, *Engines of Creation* (New York: Anchor, 1990), 81.

12. Koza, 135–137.

13. Koza, 135.

14. See Ray Kurzweil, *The Singularity Is Near: When Humans Transcend Biology* (New York: Penguin, 2005), 5 (comparing the difficulty of discovering the right magical incantation in the world of Harry Potter to the difficulty of discovering real-world incantations, "the formulas and algorithms underlying our modern-day magic").

15. See, e.g., Sadiq M. Sait and Habib Youssef, *VLSI Physical Design Automation: Theory and Practice* (River Edge, NJ: World Scientific, 1999).

16. *O'Reilly v. Morse*, 56 U.S. 62 (1854).

17. James Bessen and Robert M. Hunt, "An Empirical Look at Software Patents," *Journal of Economics and Management Strategy* 16, no. 1 (2007): 157–89.

18. See, e.g., Marc Morgan, "Stop Looking Under the Bridge for Imaginary Creatures: A Comment Examining Who Really Deserves the Title Patent Troll," *Federal Circuit Bar Journal* 17 (2007): 165–180.

19. Microsoft paid $520 million in damages as a result of a finding that it had infringed Eolas's patent. *Eolas Techs. v. Microsoft Corp.*, 2004 U.S. Dist LEXIS 522 (N.D. Ill. 2004). RIM paid NTP $612.5 million in full settlement of the claims against it. "Research in Motion and NTP Sign Definitive Settlement Agreement to End Litigation," RIM Press Release, March 3, 2006, http://www.blackberry.com/news/press/2006/pr-03_03_2006-01.shtml (accessed June 2, 2008). For other examples of

large software patent verdicts and settlements, see, e.g., Eriq Gardner, "First Bill, Now Steve," *IP Law and Business*, April 2006, http://www.burst.com/new/newsevents/articles/IP%20Law&Business.htm (accessed June 2, 2008); Lucas Graves, "Suing Your Way to Profitability," *Wired*, June 2005, 50.

Chapter 1

1. I take some literary license with my earliest experiences with computers, since my memory may blend together events that took place over the course of a year or more. I do, however, clearly remember the impact of these events on my way of thinking.

2. John Perry Barlow, "A Declaration of the Independence of Cyberspace," February 8, 1996, http://homes.eff.org/barlow/Declaration-Final.html (accessed May 31, 2008).

3. John Perry Barlow, "The Economy of Ideas," *Wired*, March 1994, http://www.wired.com/wired/archive/2.03/economy.ideas_pr.html (accessed June 2, 2008). This article covered much of the same ground as, and expanded on, the "Declaration of the Independence of Cyberspace."

4. See Nicholas Negroponte, *Being Digital* (New York: Vintage, 1996).

5. See Noel Sharkey, "I, Ropebot," *New Scientist*, July 4, 2007, 32–35. The entire description of the Ropebot offered herein is based on Sharkey's article, which also describes other early programmable machines.

6. Ibid.

7. See, e.g., Bryan Pfaffenberger, ed., *Webster's New World Computer Dictionary*, 9th Ed. (New York: Hungry Minds, 2001), 298.

8. The term *program* is even more ambiguous than this. Sometimes it is used to refer to human *ideas* about a program, or even to abstract Platonic *ideals* embodied in a program.

9. This distinction does not eliminate all possible overlap between programs and software. For example, consider a program written in a human-readable computer programming language on paper. Such a printout is also software under my definition, to the extent that you can connect a computer to a digital camera or scanner, run software on the computer to automatically capture an image of the printout, store the resulting data as software in the computer's memory, and then execute the software. The extent of this overlap, however, is no more problematic than the overlap between any of the common usages of program and software and will not cause difficulties for most of the discussion in this book.

10. See, e.g., *Microsoft Computer Dictionary*, 5th Ed. (Redmond, WA: Microsoft Press, 2002), 489, defining *software* as "instructions that make hardware work."

11. See generally, James Essinger, *Jacquard's Web: How a Hand-Loom Led to the Birth of the Information Age* (Oxford: Oxford University Press, 2004).

12. Howard Rheingold, *Tools for Thought: The History and Future of Mind-Expanding Technology* (Cambridge, MA: MIT Press, 1985), 31; Essinger, 35–36.

13. Essinger, 42.

14. In fact, it appears that the task of weaving a particular pattern into cloth using a Jacquard Loom was divided among no fewer than *three* people: a "skilled artist" who designed the pattern, a separate "peculiar artist" who used a special machine to punch holes in cards according to the pattern, and a weaver who then fed the punched cards into the loom (Essinger, 86, citing Charles Babbage's autobiography for the description of the "skilled artist" and "peculiar artist"; and 38, describing the relatively mundane work that this process relegated to the weaver who operated the loom).

15. For an overview of the birth of the Hollerith tabulator and the path from it to the founding of IBM, see generally Essinger, 149–203.

16. The Office of Charles and Ray Eames, *A Computer Perspective: Background to the Computer Age* (Cambridge, MA: Harvard University Press, 1990), 22–23; Essinger, 157–169.

17. Eames, 23; Essinger, 163–164.

18. Eames, 24–25.

19. Eames, 26; Essinger, 167–168.

20. Eames, 25.

21. Eames, 46. IBM continued to earn more revenue from punched-card systems than from computers as late as 1960 (Essinger, 172). Punched cards continued to be the primary medium for storing and then loading programs and data into computers until the 1970s. IBM manufactured its last punched-card-based machine in 1984 (Essinger, 250).

22. Eames, 123; Georges Ifrah, *The Universal History of Computing: From the Abacus to the Quantum Computer* (New York: Wiley, 2001), 212–213.

23. Eames, 49. Similarly, the early ENIAC computer was programmed by rewiring it for each problem to be solved (ibid., 138).

24. Eames, 158.

25. For background on DNA computing, see, e.g., Leonard Adleman, "Molecular Computation of Solutions to Combinatorial Problems," *Science* 266 (1994): 1021–1023; Nancy Forbes, *Imitation of Life: How Biology Is Inspiring Computing* (Cambridge, MA: MIT Press, 2004), 51–65; Moshe Sipper, *Machine Nature: The Coming Age of Bio-Inspired Computing* (New York: McGraw-Hill, 2002), 151–163.

26. See *Communications of the ACM* 50, no. 9 (September 2007). According to J. Storrs Hall, "One way to sum up nanotechnology is that it will make matter into software." J. Storrs Hall, *Nanofuture: What's Next for Nanotechnology* (Amherst: Prometheus Books, 2005), 271.

27. K. Eric Drexler claims nanoscale mechanical computers could be constructed that will be both faster and "many billions of times more compact than" the electronic computers of 1990. Drexler, 18–19.

28. Alan M. Turing recognized that whether software was stored in electrical signals was irrelevant to whether computers could be intelligent. For clues to machine intelligence, he instead suggested that we "look rather for mathematical analogies of function." Turing, "Computing Machinery and Intelligence," in *Computers and*

Thought, eds. Edward A. Feigenbaum and Julian Feldman (Menlo Park, CA: MIT Press, 1995), 16.

29. Eames, 23; Essinger, 165–166.

30. Hollerith himself recognized the limitations of such an input mechanism and eventually replaced it with keyboard-based punching devices (Essinger, 169).

31. The instruction "Add 2 + 4" is an example of "pseudocode," which resembles real computer code but is not written in a real programming language. I use it here for ease of explanation.

32. When I say the computer automatically stores the binary machine code in response to the programmer typing the source code, I am assuming, for ease of explanation, that the program is compiled automatically into machine code. Even if the machine code is generated at a subsequent time, this code is still generated automatically.

33. "1971—Microprocessor Condenses CPU Function onto a Single Chip," Computer History Museum, http://www.computerhistory.org/semiconductor/timeline/1971-MPU.html (accessed June 2, 2008).

34. Tony Fitzpatrick, "Teenager Moves Video Icons Just by Imagination," Washington University in St. Louis press release, October 9, 2006, http://news-info.wustl.edu/news/page/normal/7800.html (accessed May 31, 2008).

35. See, e.g., David Needle, "Mind/Computer Interface Advances," Internet News, November 16, 2005, http://www.internetnews.com/bus-news/print.php/3564531 (accessed May 31, 2008); Will Knight, "Virtual World Sharpens Mind-Control," NewScientist.com, June 26, 2007, http://technology.newscientist.com/article/dn12136 (accessed May 31, 2008); Richard Martin, "Mind Control," *Wired* 13, no. 3 (March 2005): 114–119.

Chapter 2

1. Never say never. What if the pattern on a cloth produced by a particular Jacquard Loom program is demonstrated to repel a certain species of bee, and therefore is useful as an insect repellant?

2. See, e.g., Turing, "Computing Machinery and Intelligence," 18–19.

3. Alan M. Turing, "On Computable Numbers with an Application to the *Entscheidungsproblem*," *Proceedings of the London Mathematical Society* 42 (1936): 230–267.

4. Hilbert had asked whether mathematics was *decidable*—in other words, whether there existed a "definite method" that could be used to determine whether *any* mathematical statement was true or false. Turing proved that there was no such method. Andrew Hodges, *Alan Turing: The Enigma* (New York: Walker, 2000), 91–104.

5. John von Neumann, "First Draft of a Report on the EDVAC," Contract No. W-670-ORD-4926, between the U.S. Army Ordnance Department and the University of Pennsylvania. Moore School of Electrical Engineering, University of Pennsylvania, June 30, 1945.

6. Rheingold, 66.

7. The design for Babbage's Analytical Engine called for it to store both program and data on punched cards (Essinger, 89).

8. Eames, 137.

9. Storing program and data in a single memory of a universal Turing machine was always possible in theory. See, e.g., Turing, "Computing Machinery and Intelligence."

10. To exclude a physical device from qualifying as an invention merely because the device can be described using mathematical language would be to create an exception that would swallow the rule, since all physical devices are susceptible of description in mathematical terms. Judge Pauline Newman acknowledged as such when she recognized that describing software in mathematical terms does not make it into an unpatentable "abstract idea": "Mathematics is not a monster to be struck down or out of the patent system, but simply another resource whereby technological advance is achieved." *In re Alappat*, 33 F.3d 1526 (Fed. Cir. 1994. Newman, J., dissenting).

11. See Chapter 7.

12. I don't claim that software controlling external devices is the *only* kind of software that can qualify as an invention. For example, software that organizes the memory of a computer efficiently might be an invention. See *In re Lowry*, 32 F.3d 1579 (Fed. Cir. 1994). I focus on software for controlling external devices simply because it is the class of software that most easily fits within the definition of invention.

13. The field of Cybernetics, which strongly influenced the development of early computers, is the study of control and communication in natural and artificial systems. See Norbert Weiner, *Cybernetics: Or the Control and Communication in the Animal and the Machine*, 2nd Ed. (Cambridge, MA: MIT Press, 1965).

14. The term *mechatronics* refers to the integration of traditional mechanical systems with electronic components controlled by software. See Rolf Isermann, "Mechatronics," *MIT Technology Review*, February 2003, 40–41.

15. Dartmouth Philosophy Professor James Moor coined the term *logical malleability* to refer to the ability of computers to be manipulated by changing the logical instructions given them. James H. Moor, "The Future of Computer Ethics: You Ain't Seen Nothin' Yet!" *Ethics and Information Technology* 3, no. 2 (2001): 89–91.

16. See, e.g., Rheingold, 14; Paul Armer, "Attitudes Toward Intelligent Machines," reprinted in Feigenbaum and Feldman, 389–405.

17. Paul Ceruzzi, *Reckoners: The Prehistory of the Digital Computer, from Relays to the Stored Program Concept* (Westport, CT: Greenwood Press 1983), 43.

18. Rheingold, p. 86.

19. See Betty Alexandra Toole, ed., *Ada, the Enchantress of Numbers* (Mill Valley, CA: Strawberry Press, 1992); see also Rheingold, 25, 31.

20. Ada Byron is referred to in the literature alternatively as Augusta Ada Byron, Ada Byron, Ada Lovelace, Lady Lovelace, and perhaps most frequently simply as Ada, perhaps to avoid the confusion that would arise with her more famous father if she were referred to simply as Byron in the way that we refer to Babbage, Turing, and others.

21. Rheingold, 28. Also, see Toole. Ada's most-cited work is the set of notes that she wrote as an appendix to her translation into English of Italian scientist Luigi Federico Menabrea's paper discussing Babbage's Analytical Engine. Ada's notes, which have far surpassed in notoriety the paper on which they comment, are three times as long as Menabrea's paper itself. See Essinger, 122, 132–145.

22. Rheingold, 30–34, citing Philip Morrison and Emily Morrison, eds., *Charles Babbage and His Calculating Engines* (New York: Dover, 1961), 251–252; Essinger, 81–98. Although a full working version of the Analytical Engine has never been constructed, Babbage's son built part of the engine after Babbage's death (Rheingold, 30).

23. Essinger, 140–141, citing Ada's Notes (emphasis in original).

24. Rheingold, 33–34, citing Doris Langley Moore, *Ada, Countess of Lovelace: Byron's Legitimate Daughter* (New York: Harper and Row, 1977), 155.

25. See Rheingold, 48.

Chapter 3

1. For a general overview of the role of abstraction in computer science, see Jeff Kramer, "Is Abstraction the Key to Computing?" *Communications of the ACM* 50, no. 4 (April 2007): 37–42.

2. Edsger W. Dijkstra, "The Humble Programmer," *Communications of the ACM* 15, no. 10 (October 1972): 859–866.

3. This process of building abstractions is so fundamental to computer science that the first two chapters of the introductory computer science text used at MIT are titled "Building Abstractions with Procedures" and "Building Abstractions with Data." Harold Abelson, Gerald Jay Sussman, and Julie Sussman, *Structure and Interpretation of Computer Programs* (Cambridge, MA: MIT Press, 1985): 1–166.

4. Alan Turing recognized early on the possibility of building hierarchies of automatically translatable abstractions (Rheingold, 61).

5. For example, in 1936 Turing used the hypothetical example "of a man in the process of computing a real number" as a "computer." Turing, "On Computable Numbers." Even earlier, the clerks who performed arithmetical calculations necessary to create astronomical and mathematical tables in Babbage's day were referred to as computers (Essinger, 66).

6. For a history of such unsung "human computers" in the early days of computing, see David Alan Grier, *When Computers Were Human* (Princeton: Princeton University Press, 2005).

7. See Edsger Dijkstra, "A Preliminary Investigation into Computer Assisted Programming" (EWD237), http://www.cs.utexas.edu/EWD/transcriptions/EWD02xx/EWD237.html (accessed May 31, 2008).

8. Martin Campbell-Kelly, *From Airline Reservations to Sonic the Hedgehog: A History of the Software Industry* (Cambridge, MA: MIT Press, 2003), 34.

9. Mathematician Grace Hopper wrote the first practical software compiler for the Eckert-Mauchly Computer Corporation (Eames, 135). For a discussion of early

recognition of the need for and actual development of compilers, see Thierry Bardini, *Bootstrapping: Douglas Engelbart, Coevolution, and the Origins of Personal Computing* (Stanford, CA: Stanford University Press, 2000), 2–3. Not all high-level languages require compilers to make their programs run. Some languages use *interpreters*, which translate each individual instruction right before it is executed, rather than translating all instructions ahead of time. In either case, however, instructions in high-level languages must be translated into machine language before they can be executed.

10. Similarly, the meaning of the term *automatic programming* continuously changes as it becomes possible to automatically translate instructions written at increasingly high levels of abstraction into machine language. In response to this fluidity of meaning, computer scientist David Parnas concluded that "automatic programming has always been a euphemism for programming in a higher-level language than was then available to the programmer." D. L. Parnas, "Software Aspects of Strategic Defense Systems," *Communications of the ACM* 28, no. 12 (November 1985): 1326–1335.

11. Libraries are one way of enabling subroutines to be reused so that subsequent programmers do not have to reinvent the wheel. For information about reuse more generally, see Hafedh Mili et al., *Reuse-Based Software Engineering: Techniques, Organization, and Controls* (New York: Wiley, 2002).

12. Essinger, 91; Brad J. Cox, "Planning the Software Industrial Revolution," *IEEE Software*, November 1990, 25–33.

13. Programs have also become more abstract over time as a result of hardware "virtualization." Suppose a program written in an early high-level programming language included an instruction to store a number in memory location number 234. Modern program instructions do not typically refer directly to specific physical memory locations in this way. Instead, a program instruction referring to memory location 234 is translated automatically to refer to another memory location, just as the postal service delivers an envelope addressed to the "Returns Department" to the appropriate warehouse at a mail order facility. As a result of such automatic translation, the programmer does not have to know anything about the hardware on which the program will run, and the same program can run on many computers with dissimilar memory configurations. For more examples of hardware virtualization, see, e.g., Jon Crowcroft, "On the Nature of Computing," *Communications of the ACM* 48, no. 2 (February 2005): 19–20.

14. U.S. Patent No. 6,016,394 (issued January 18, 2000).

15. Jeff Walker, interview by author, March 9, 2006.

16. Campbell-Kelly, 34–35.

17. Carver Mead and Lynn Conway, *Introduction to VLSI Systems* (Reading, MA: Addison-Wesley, 1980), 396.

18. *O'Reilly* v. *Morse*, 56 U.S. 62, 94 (1854) ("very soon after the discovery made by Oersted [of the electro-motive power], it was believed by men of science that this newly-discovered power might be used to communicate intelligence to distant places"). Similarly, many in Jacquard's day recognized the potential benefits of an improved drawloom and were working on inventing one (Essinger, 17).

19. Richard Stallman, "For Submission to the Patent & Trademark Office," January 1994, http://lpf.ai.mit.edu/Patents/rms-pto.html (accessed May 31, 2008). See also Richard Stallman and Simson Garfinkel, "Against Software Patents," *Communications of the ACM* 35, no. 1 (January 1992): 17–22, 121 ("A computer program is built from ideal mathematical objects").

20. Jim Warren, "Autodesk Statement on Software Patents," January 26–27, 1994, http://www.jamesshuggins.com/h/tek1/software_patent_autodesk.htm (accessed May 31, 2008).

21. Donald Knuth, "Letter to the Patent Office from Professor Donald Knuth," 1994, http://progfree.org/Patents/knuth-to-pto.txt (accessed May 31, 2008).

22. Yet another claim along these lines is that software is merely a mathematical object, and that software therefore should not be patentable because advances in pure mathematics are not patentable. See, e.g., Ben Klemens, *Math You Can't Use: Patents, Copyright, and Software* (Washington, D.C.: Brookings Institution Press, 2006), 26–27, 44–45.

23. Ifrah, 120, 123–124.

24. For similar arguments, see for example John Searle, *Minds, Brains and Science* (Boston: Harvard University Press, 1984), 48–49; Dan L. Burk, "Patenting Speech," *Texas Law Review* 79 (2000): 99–135.

25. R. W. Hamming, "One Man's View of Computer Science," *Journal of the Association for Computing Machinery* 16 (1969): 3, 5.

26. Agatha Christie, *An Autobiography* (New York: Dodd, Mead, 1977).

27. See Rheingold, 38.

28. Eames, 34.

29. Rheingold, 27–28, citing Philip Morrison and Emily Morrison, eds., *Charles Babbage and His Calculating Engines* (New York: Dover Publications, 1961), 33; see also Anthony Hyman, *Charles Babbage: Pioneer of the Computer*, (Princeton: Princeton University Press, 1982), 49–50.

30. Rheingold, 28.

31. Ifrah, 101, quoting L. F. Menabrea, "Sur la Machine Analytique de Charles Babbage," *Comptes rendus des séances de l' Académie des sciences Paris*, 28 July 1884, 179–182.

32. Rheingold, 132–135.

33. Even before the term *software* existed, Turing noted that programming would always remain interesting because any aspect of it that became routine and therefore boring could be automated by machine (Hodges, 396).

34. See Goldberg, *The Design of Innovation*, 228.

Chapter 4

1. Although I refer to the Creativity Machine as software, it is actually a paradigm that can be implemented in hardware or software, much as the von Neumann architecture can be implemented in a variety of hardware. Stephen Thaler, interview

by author, May 12, 2005. Imagination Engines has obtained a variety of patents on the Creativity Machine and other technologies. See, e.g., U.S. Patent No. 5,659,666 (issued August 19, 1997); U.S. Patent No. 6,018,727 (issued January 25, 2000); U.S. Patent No. 6,356,884 (issued March 12, 2002).

2. Stephen Thaler, email message to author, February 17, 2008, and interview by author, February 20, 2008.

3. Stephen Thaler, email message to author, February 17, 2008, and interview by author, February 20, 2008.

4. Stephen Thaler, email message to author, February 17, 2008, and interview by author, February 20, 2008.

5. The Creativity Machine produced not one but about 2,000 potentially superior toothbrush designs, many of which had crossed bristles. Dr. Thaler turned all of these designs over to Gillette, which made the final choice and may have made modifications for use in the CrossAction toothbrush (Stephen Thaler, email message to author, May 13, 2008).

6. For overviews of how traditional artificial neural networks work, see Forbes, 4–8; Sipper, 51–63; Kurzweil, 268–270.

7. See, e.g., Forbes, 8–9; Sipper, 64; David B. Fogel, *Evolutionary Computation: Toward a New Philosophy of Machine Intelligence*, 2nd Ed. (New York: IEEE Press, 2000), 16–19.

8. University of Sussex Professor Margaret Boden has identified the ability to produce ideas or artifacts that are surprising as an element of creativity. According to Boden, an idea may be surprising because (1) it is unfamiliar or unlikely; (2) you are surprised to find that the idea fits into an existing framework; or (3) you previously considered the idea to be impossible. Margaret A. Boden, *The Creative Mind: Myths and Mechanisms*, 2nd Ed. (New York: Routledge, 2004), 1–3.

9. For examples of Koza's work, see *Genetic Programming IV*; W. Wayt Gibbs, "Programming with Primordial Ooze," *Scientific American*, October 1996; John R. Koza, Martin A. Keane, and Matthew J. Streeter, "Evolving Inventions," *Scientific American*, February 2003, 52–59. The concept of genetic programming was first proposed by Nichael Cramer in 1985 (Sipper, 34).

10. See Koza, *Genetic Programming IV*, 12 (characterizing the fitness function as "the primary mechanism for communicating the high-level statement of the problem's requirements to the genetic programming system").

11. Koza has published a series of books on genetic programming, documenting the technology's increasingly powerful abilities as an automated invention machine. In his most recent *tour de force*, *Genetic Programming IV: Routine Human-Competitive Machine Intelligence*, he documents 36 instances in which he and his team have used genetic programming to produce problem solutions that compete with or exceed those produced by human experts. Koza also has patented a variety of techniques for using genetic programming and genetic algorithms more generally. See, e.g., U.S. Patent No. 6,424,959 (issued July 23, 2002); U.S. Patent No. 6,058,385 (issued May 2, 2000); U.S. Patent No. 5,390,282 (issued February 14, 1995).

12. The controller is patented: U.S. Patent No. 6,847,851 (issued January 25, 2005). A method of designing the controller and other controllers is patented: U.S. Patent No. 7,117,186 (issued October 3, 2006).

13. Koza, *Genetic Programming IV*, 50 and sources cited therein.

14. Ibid., 56.

15. Ibid., 57.

16. Ibid., 57–58.

17. Ibid., 87, referring to the PID controller design of Dorf and Bishop.

18. For more information on evolutionary computation, see generally Peter J. Bentley, *Digital Biology* (New York: Simon & Schuster, 2001), 40–64; Sipper, 13–20.

19. For a brief overview of the operation and history of genetic algorithms, see Melanie Mitchell, *An Introduction to Genetic Algorithms* (Cambridge, MA: MIT Press, 1998); Fogel, 73–82.

20. Sipper, 193.

21. The complete set of requirements embodied in Koza's controller for a fitness function can be found in Koza, *Genetic Programming IV*, 87–90.

22. Ibid., 87.

23. Ibid., 87.

24. Hornby et al., "Computer-Automated Evolution."

25. For the complete fitness function, see Jason D. Lohn, Gregory Hornby, and Derek Linden, "An Evolved Antenna for Deployment on NASA's Space Technology 5 Mission," in *Genetic Programming Theory and Practice II*, eds. Una-May O'Reilly et al. (New York: Springer, 2006), 301–315.

26. David Fogel, interview by author, September 20, 2007.

27. Daniel Riester et al., "Thrombin Inhibitors Identified by Computer-Assisted Multiparameter Design," *PNAS* 102, no. 24 (June 14, 2005): 8597-8602; Sion Balass, email message to author, March 19, 2008.

28. Icosystem Corporation, "The Hunch Engine: Molecule Search for Drug Discovery," Icosystem, icosystem.com/hunch_mobius.htm (accessed May 31, 2008).

29. Takenori Wajima, "Latest System Technologies for Railway Electric Cars," Hitachi, Ltd., http://www.hitachi.com/ICSFiles/afieldfile/2005/12/26/r2005_04_102.pdf (accessed May 31, 2008).

30. Brian Mattmiller, "Building Better Engines Through Natural Selection," *Foundry Management and Technology*, October 2001, 2.

31. "Maxygen's Next-Generation Interferon Alpha Enters Phase Ia Clinical Trial," Maxygen, November 7, 2006, http://www.maxygen.com/newsview.php?listid=263 (accessed May 31, 2008).

32. For a history of genetic algorithms and evolutionary computation more generally, see David B. Fogel, ed., *Evolutionary Computation: The Fossil Record* (New York: Wiley-IEEE Press, 1998). For examples of other computer technologies that are based on analogies to biology, see Forbes. Alan Turing anticipated the basic features of evolutionary computation in Turing, "Computing Machinery and Intelligence."

33. Gregory S. Hornby and Tina Yu, "Results of the First Survey of Practitioners of Evolutionary Computation," *ACM SIGEVOlution* 2, no. 1 (Spring 2007): 2–8.

Chapter 5

1. Genetic programming is by no means the only technology that can be used to write software automatically. See, e.g., Sandra Upson, "Computer Software That Writes Itself," *Newsweek*, January 2, 2006, http://www.newsweek.com/id/51529 (accessed June 2, 2008).

2. John F. Jacobs of Lincoln Laboratory generally is credited with introducing the waterfall model to the software industry in a talk titled "The Romance of Programming," in November 1956. By 1970, this technique and similar ones were standard practice in the software industry (Campbell-Kelly, 67–69).

There is nothing special about the waterfall model in relation to invention automation; I use it only as one example of many models of engineering design that could support the same conclusions equally well. For overviews of the waterfall model as it applies to software, see Shari Lawrence Pfleeger, *Software Engineering: Theory and Practice*, 2nd Ed. (Upper Saddle River, NJ: Prentice Hall, 2001), 21–27, 48–51; Dick Hamlet and Joe Maybee, *The Engineering of Software: Technical Foundations for the Individual* (Boston: Addison-Wesley, 2001), 87–91; James F. Peters and Witold Pedrycz, *Software Engineering: An Engineering Approach* (New York: Wiley, 2000), 45–49.

3. For an overview of requirements analysis, see, e.g., Peters et al., 117–137.

4. The Wright Brothers used rigorous functional decomposition to invent the airplane. Goldberg, *The Design of Innovation*, 12–13; Orville Wright, *How We Invented the Airplane*, ed. Fred C. Kelly (New York: Dover, 1988), v.

5. Although chemists traditionally have invented chemicals without "designing" their physical structure, they still must perform chemical experiments and ascertain that the physical chemical they have produced solves the problem at hand. In such instances, the human chemist is still responsible for *identifying*, if not designing, a physical chemical that solves the problem.

6. *Cameron & Everett v. Brick*, 1871 Dec. Comm'r Pat. 89, 90 (Apr. 1, 1871).

7. "The programmer's activity ends when he has constructed a correct program," since "once the program has been made, the 'making' of the corresponding process is delegated to the machine." Edsger Dijkstra, "Go To Statement Considered Harmful," reprinted in *Communications of the ACM* 51, no. 1 (January 2008): 7–8.

8. A fitness function and the other initial conditions required for genetic programming might be best characterized either as a problem statement or as a set of requirements, but which of these is more accurate is irrelevant to our purposes, since both are more abstract than the functional specification represented by a traditional computer program.

9. See, e.g., Irving John Good, "Speculations Concerning the First Ultraintelligent Machine," *Advances in Computers* 6 (1965): 31–88, quoted in Kurzweil, 22.

10. This class of procedure is sometimes referred to as a "population-oriented" or

"selectionist" approach, since it relies on generating and selecting from among populations of potential solutions (David Goldberg, interview by author, July 28, 2005).

11. For a similar conclusion, see Herbert A. Simon, *The Sciences of the Artificial,* 3rd ed. (Cambridge, MA: MIT Press, 1996), 195.

12. A. L. Samuel, "Some Studies in Machine Learning Using the Game of Checkers," in Feigenbaum and Feldman, 71–105.

13. Goldberg, *The Design of Innovation,* 17, 52.

14. David Goldberg, interview by author, July 28, 2005. See also Simon, 120, 128, 190–193.

15. Rheingold, 211; M. Mitchell Waldrop, *The Dream Machine: J.C.R. Licklider and the Revolution That Made Computing Personal* (New York: Penguin, 2002), 143.

16. Rheingold, 210.

17. Rheingold, 211.

18. Rheingold, 217.

19. See Mihály Csíkszentmihályi, *Creativity: Flow and the Psychology of Discovery and Invention* (New York: Harper Perennial, 1996).

20. See, e.g., Chuan-Hoo Tan and Hock-Hai Teo, "Training Future Software Developers to Acquire Agile Development Skills," *Communications of the ACM* 50, no. 12 (December 2007): 97–98; Dror G. Feitelson, "Experimental Computer Science," *Communications of the ACM* 50, no. 11 (November 2007): 24–26.

21. Similarly, the giant in computer-aided design (CAD), Autodesk, recently announced a relatively inexpensive ($5,300) software package called "Inventor" that can perform 3D virtual prototyping, which Autodesk claims can be used to test how the different elements of a design will respond to gravity or torque, thereby eliminating the need to build and test physical prototypes. Mark Borden et al., "The World's Most Innovative Companies," *Fast Company,* March 2008, 104.

22. See, e.g., Hall, 71–72.

23. Jeff Walker, interview by author, March 9, 2006.

24. For a brief overview of structured programming, see Robert W. Floyd, "The Paradigms of Programming," *Communications of the ACM* 22, no. 8 (August 1979): 455–460.

25. Campbell-Kelly, 67–69.

26. "Illinois Researchers Break Billion Variable Optimization Barrier," University of Illinois College of Engineering, January 19, 2007, http://www.engr.uiuc.edu/news/index.php?xId=070208160798 (accessed May 31, 2008).

27. Dijkstra, "The Humble Programmer."

28. For a summary of the decrease in size of machines over time, see Kurzweil, 82–84.

29. Mark Weiser first introduced the concept of pervasive computing in "The Computer for the 21st Century," *Scientific American,* September 1991, 94–104. For an overview of pervasive computing, see Debashis Saha and Amitava Mukherjee, "Pervasive Computing: A Paradigm for the 21st Century," *IEEE Computer,* March 2003, 25–31.

30. Weiser, "The Computer for the 21st Century."

31. Mark Weiser, "The World Is Not a Desktop," *ACM Interactions* 1, no. 1 (January 1994), 7–8.

32. Goldberg, *The Design of Innovation*, 228.

33. For a variety of data on this and other exponential increases in features of computer technology, see Kurzweil, 47–84.

34. "Multiple realizability" can also refer to the ability of a single algorithm to be implemented in numerous computer programs. Generally, a description at *any* level of the waterfall is multiply realizable if it can be implemented by more than one description at a lower level of the waterfall.

35. Campbell-Kelly, 98.

36. Evolutionary computation was useful for designing the NASA antenna we encountered earlier because we do not have any clear-cut rules for antenna design; rather, designing antennas is an art. Since, however, we are able to *simulate* the operation of antennas accurately, evolutionary algorithms were a good way to design them. Gregory Hornby, interview by author, June 28, 2005.

37. See, e.g., Daniel H. Pink, *A Whole New Mind: Moving from the Information Age to the Conceptual Age* (New York: Riverhead Books, 2005), 41–44.

38. See David B. Fogel, *Evolutionary Computation: Toward a New Philosophy of Machine Intelligence*, 2nd Ed. (New York: IEEE Press, 2000), 8–9.

39. Steven Johnson, *Emergence* (New York: Scribner, 2001), 170–174, citing W. Daniel Hillis, *The Pattern on the Stone: The Simple Ideas That Make Computers Work* (New York: Basic Books, 1999).

40. For example, Joe Rothermich of Natural Selection told me that some of his company's clients require solutions they can understand for use in mission-critical situations, or to satisfy applicable legal requirements. Joe Rothermich, interview by author, September 20, 2007.

41. See Sipper, 45.

42. Gibbs, 30–31.

43. Marcel Theurk and Sion Balass, interview by author, November 15, 2007; Sion Balass, interview by author, October 26, 2005. See also Gibbs.

44. University of Idaho Professor James Foster told me that "the key difference [between evolutionary computation and classical engineering] is that [the former] is a technology that allows your creations to surprise you." James Foster, interview by author, June 27, 2005. See also Sipper, 197.

45. Gregory Hornby, interview by author, June 28, 2005.

46. Alan Turing called this "Lady Lovelace's objection" to the assertion that machines could be intelligent, in reference to Ada Byron's elucidation of it in 1842. Turing rejected the objection on the grounds that programmers cannot necessarily predict the "remote consequences" of the programs they write. Turing, "Computing Machinery and Intelligence," in Feigenbaum and Feldman, 26–27.

47. The primary difference between today's evolutionary computation and more traditional forms of artificial intelligence such as expert systems is that today the focus

is on producing systems that discover the rules to use in solving problems, rather than programming those rules directly into the system. John Holland, interview by author, September 27, 2005. See Johnson, *Emergence.*

48. See Gibbs.

49. Cem Baydar, interview by author, June 27, 2005. Genetic algorithms can also be used to address problems having a larger number of variables than traditional methods, such as linear programming, because in such situations the genetic algorithm is likely to be able to find at least *some* improvement over previous solutions even on very complex problems. More generally, genetic algorithms are applicable to noisy, large, and ill-defined problems because they are not limited to yielding an all-or-nothing solution. David Goldberg, interview by author, July 28, 2005. Similarly, evolutionary computation is useful in situations in which we lack a definition of what is optimal, or in which we would be satisfied with a solution that merely constitutes an improvement over existing solutions, even if the improvement is not optimal. John Holland, interview by author, September 27, 2005.

50. Frank Lewis Dyer and Thomas Commerford Martin, *Edison, His Life and Inventions* (New York: Harper and Brothers, 1929), 234–266. See also Ronald W. Clark, *Edison, the Man Who Made the Future* (New York: Putnam, 1977), 87.

51. Edison alone held a record 1,093 U.S. patents. A full listing of patents issued in Edison's name is available at http://www.tomedison.org/patent.html (accessed June 2, 2008).

52. U.S. Const. Art. 1, Sec. 8, Cl. 8; 35 U.S.C. § 100.

53. Feigenbaum and Feldman, 3. See also Arthur Samuel, "AI: Where It Has Been and Where It Is Going," *Proceedings of the 8th International Joint Conference on Artificial Intelligence* (1983): 1152–1157.

54. Peter Bentley, email message to author, November 10, 2006.

55. Such skepticism may also arise from the same cause as skepticism toward Darwinian evolution: our tendency to assume that "it takes a big fancy smart thing to make a lesser thing. . . . You'll never see a spear making a spear maker." Daniel Dennett, "Darwinism Completely Refutes Intelligent Design," interview, *Der Spiegel*, December 26, 2005, http://www.spiegel.de/international/spiegel/0,1518,392319,00.html (accessed June 2, 2008). The same assumption might lead one to conclude that no search-based process, acting according to simple rules, could ever produce a complex machine.

56. Stephen Thaler believes that one of the primary obstacles standing in the way of accepting the Creativity Machine is "all too human pride." He "predict[s] that this vanity will actually stand in the way of human progress." Stephen Thaler, "Interview with Atomasoft Reporter Sander Olson," http://www.imagination-engines.com/atomasoft.htm (accessed June 2, 2008).

57. See, e.g., Sipper, 47.

58. On a similar note, James Foster told me that "you have to be a lot more relaxed in this area [evolutionary computation], because you know you're going to get an answer that you don't understand, and the answer is probably going to surprise

you. You just have to be comfortable with that. That is not a classical engineering personality trait." James Foster, interview by author, June 27, 2005.

59. See interview with Michael Conrad of Wayne State University in Janine M. Benyus, *Biomimicry: Innovation Inspired by Nature* (New York: HarperCollins, 1997), 211–212, for more on his view that we "may have to loosen up the reins a bit" if "we are to break out of our control-hungry straitjacket and achieve true power."

Chapter 6

1. See Clive Thompson, "Your Outboard Brain Knows All," *Wired*, September 25, 2007, 66.

2. Kurzweil, 305.

3. Joel Garreau, *Radical Evolution: The Promise and Peril of Enhancing Our Minds, Our Bodies—and What It Means to Be Human* (New York: Doubleday, 2005), 27.

4. Andy Clark, *Natural Born Cyborgs* (New York: Oxford University Press, 2003), 16–18.

5. Clark, 4.

6. Clark, 24.

7. See ibid. for an excellent exposition of the claim that humans' true cyborg nature lies not necessarily in physical implantation of artificial devices in the human body but in integration of human thought processes with machines, even when those machines remain external to the human body.

8. Clark, 24. J.C.R. Licklider, "Man-Computer Symbiosis," *IRE Transactions on Human Factors in Electronics* HFE-1 (March 1960), 4.

9. Clark, 24, and sources cited in Clark, 24n19.

10. See generally, Douglas C. Engelbart, "A Conceptual Framework for the Augmentation of Man's Intellect," in *Vistas in Information Handling*, eds. Paul William Howerton and David C. Weeks (Washington, D.C.: Spartan Books, 1963), 1–29. For an overview of Engelbart's work, and an attempt to explain why much of his work has been overlooked or misunderstood, see Bardini.

11. Engelbart, "A Conceptual Framework," 4–5.

12. Bardini, 11, citing Ross Ashby, "Design for an Intelligence Amplifier," in *Automata Studies*, eds. C. E. Shannon and J. McCarthy (Princeton: Princeton University Press, 1956), 216.

13. See generally, Engelbart, "A Conceptual Framework."

14. Ibid., 4–5.

15. Rheingold, 183–184.

16. Engelbart, "A Conceptual Framework," 6.

17. Clark, 6, presents a similar example of performing multiplication longhand on paper. Douglas Engelbart realized that when we humans attempt to deal with more than a few concepts at a time, we require some external medium in which to store those concepts while we process them. Bardini, 41, quoting Douglas C. Engelbart,

"Special Considerations of the Individual as a User, Generator, and Retriever of Information," *American Documentation* 12, no. 2 (1961): 122.

18. Gregory Hornby, interview by author, June 28, 2005, and email message to author, March 6, 2008.

19. James Foster, interview by author, June 27, 2005.

20. Hornby, one of the NASA engineers responsible for creating the antenna we saw in the Introduction, believes that evolutionary computation "doesn't put people out of work; it just makes them more useful. It actually is a tool for making people more efficient." Gregory Hornby, interview by author, June 28, 2005.

21. Licklider thought that such a partnership was the "best arrangement" for both human and computer (Rheingold, 140).

22. Rheingold, 174. Even earlier, von Neumann recognized that the general-purpose machine he was working on in the form of the ENIAC had all the makings of a mind extension tool (ibid., 86).

23. Goldberg, *The Design of Innovation*, 228.

24. Babbage, Herman Goldstine (one of the creators of the ENIAC), Turing, and Licklider are all further examples of innovators who invented new machines to assist themselves in solving problems (Rheingold, 142).

25. Campbell-Kelly, 100–101.

26. Bardini, 24, 56–57.

27. Jeff Walker, interview by author, March 9, 2006.

28. Bardini, 56–57.

Chapter 7

1. See 35 U.S.C. § 154(a).

2. U.S. Const. art. I, § 8, cl. 8.

3. 35 U.S.C. § 154(a)(2).

4. See Pamela Samuelson et al., "A Manifesto Concerning the Legal Protection of Computer Programs," *Columbia Law Review* 94 (1994): 2308–2431, 2365–2371.

5. See, e.g., *Pennock* v. *Dialogue*, 27 U.S. 1, 23 (1829); *Universal Oil Prods. Co.* v. *Globe Oil & Refining Co.*, 332 U.S. 471 (1944).

6. 35 U.S.C. § 102.

7. The Supreme Court recently reaffirmed the importance of prohibiting patents on inventions that would have been released into the stream of commerce without patent protection. "Granting patent protection to advances that would occur in the ordinary course without real innovation retards progress." *KSR Int'l, Co.* v. *Teleflex, Inc.*, 127 S.Ct. 1727 (2007).

8. Furthermore, other mechanisms, such as grants and contests can produce innovation in the absence of patents. For example, both margarine and the process of canning food were invented in response to contests. Anya Kamenetz, "The Power of the Prize," *Fast Company*, May 2008, 43–45. For a general discussion of forms of

incentives, including patents, grants, and contests, see Suzanne Scotchmer, *Innovation and Incentives* (Cambridge, MA: MIT Press, 2004), 1–29.

9. *Graham* v. *John Deere*, 383 U.S. 1, 11 (1966).

10. It is not possible to give general examples of what kinds of modifications constitute nonobvious modifications; the determination must always be made in light of the particular facts of the case, especially the state of the art at the time the invention was made. This is part of what makes the nonobviousness requirement so slippery. Even lengthening a car frame by an inch might not be obvious, if doing so has some useful and unexpected benefit (such as making it significantly more aerodynamic).

11. As the Supreme Court has stated, "If a person of ordinary skill can implement a predictable variation, § 103 likely bars its patentability. For the same reason, if a technique has been used to improve one device, and a person of ordinary skill in the art would recognize that it would improve similar devices in the same way, using the technique is obvious unless its actual application is beyond his or her skill." *KSR Int'l, Co.* v. *Teleflex, Inc.*, 127 S.Ct. 1727 (2007).

12. *Diamond* v. *Chakrabarty*, 447 U.S. 303, 309 (1980).

13. See generally *State Street Bank & Trust Co.* v. *Signature Financial Group, Inc.*, 149 F.3d 1368 (Fed. Cir. 1998).

14. 35 U.S.C. § 112 ¶1.

15. 35 U.S.C. § 112 ¶2.

16. The courts have explicitly recognized that one way in which the patent system is intended to promote innovation is by encouraging inventors to design around the patents of their competitors. See, e.g., *Festo Corp.* v. *Shoketsu Kinzoku Kogyo Kabushiki Co., Ltd.*, 535 U.S. 722, 727 (2002). For a thorough discussion of designing around patents in the software industry, see Julie E. Cohen and Mark A. Lemley, "Patent Scope and Innovation in the Software Industry," *California Law Review* 89 (2001): 1–57.

17. *United States* v. *Dubilier Condenser Corp.*, 289 U.S. 178, 186–187 (1933).

18. See 35 U.S.C. § 271, which defines the acts that constitute patent infringement.

19. Eliezer S. Yudkowsky, "Staring into the Singularity," November 18, 1996 (updated May 27, 2001), http://sysopmind.com/singularity.html (accessed June 2, 2008).

20. See Dan L. Burk and Mark A. Lemley, "Is Patent Law Technology-Specific?" *Berkeley Technology Law Journal* 17 (2002): 1155–1206, 1179, 1193–94.

21. See, e.g., *KSR Int'l, Co.* v. *Teleflex, Inc.*, 127 S.Ct. 1727 (2007).

22. Although I describe this throughout as a "formula," it could be any abstract description in the language of mathematics or physics, an algorithm describing efficient airflow, or a fitness function or other set of abstract criteria that an aerodynamic car frame must satisfy.

23. The Supreme Court has recognized that the advancing state of the art informs the obviousness analysis: "We build and create by bringing to the tangible and palpable reality around us new works based on instinct, simple logic, ordinary inferences, extraordinary ideas, and sometimes even genius. These advances, once part of our shared knowledge, define a new threshold from which innovation starts once more. And as progress beginning from higher levels of achievement is expected in the nor-

mal course, the results of ordinary innovation are not the subject of exclusive rights under the patent laws. Were it otherwise patents might stifle, rather than promote, the progress of useful arts." *KSR Int'l, Co. v. Teleflex, Inc.*, 127 S.Ct. 1727 (2007).

24. This and other statements about the existing interpretation of the nonobviousness requirement are limited to U.S. law.

25. The failure of the courts to confirm expressly that the nonobviousness analysis should take into account the effects of technological advancement on the inventor's skill level is particularly discouraging in light of the specific invitation to do so that was recently extended to the Supreme Court. In an *amicus curiae* brief written by Professor Katherine Strandburg and signed by fourteen of our nation's most influential intellectual property law professors, Strandburg suggested that the Court acknowledge that "skill" comprises the "routine experimentation and application of ordinary tools, methods, and problem-solving abilities." Brief of Intellectual Property Law Professors as *Amici Curiae* in Support of Petitioner at 14, *KSR Inter. Co. v. Teleflex, Inc.*, 127 S. Ct. 1727 (2007). Unfortunately, the Supreme Court's decision in the *KSR* case did not clearly accept this invitation, leaving the state of the law unclear on this point.

26. There might be situations in which you are required to disclose your use of artificial invention technology, or in which it would benefit you to do so. For example, doing so might help justify a broad patent claim covering multiple car frame designs that could be generated by artificial invention technology on the basis of the same airflow equation.

27. 35 U.S.C. § 103(a).

28. Only "unobvious developments which would not occur spontaneously from the application of . . . ordinary skill" are intended to be patentable. Giles S. Rich, "The Principles of Patentability," *Journal of the Patent Office Society* 42 (1960): 75, 81-82.

29. Similarly, the early introduction of electronic calculators into businesses put skilled human calculators out of work because "ordinary workers, equipped with machines, could do the job" (Eames, 34).

30. Engelbart, "A Conceptual Framework," 5.

31. See Michael J. Meurer, "Business Method Patents and Patent Floods," *Washington University Journal of Law and Policy* 8 (2002): 309–339.

32. *In re Winslow*, 365 F.2d 1017, 151 USPQ 48 (CCPA 1966).

33. See, e.g., Jan A. Bergstra and Paul Klint, "About 'Trivial' Software Patents: The IsNot Case," *Science of Computer Programming* 64, no. 3 (February 2007): 264–285; Stallman and Garfinkel; Klemens, 1–2.

34. See Jeffrey D. Ullman, "Ordinary Skill in the Art," November 16, 2000 (minor updates August 30, 2001), http://infolab.stanford.edu/ullman/pub/focs00.html (accessed May 30, 2008).

35. Indeed, another fault in our 19th-century mind-set when it comes to the nonobviousness doctrine is that the courts overemphasize the *level of education* of the person of ordinary skill and frequently equate education with skill. In the *KSR* case, for example, "the District Court determined . . . that *the level of ordinary skill* in pedal design was 'an *undergraduate degree* in mechanical engineering (or the equivalent

amount of industry experience) [and] *familiarity with pedal control systems* for vehicles'" (emphasis added). *KSR Int'l, Co. v. Teleflex, Inc.*, 127 S.Ct. 1727 (2007). The Supreme Court did not contradict this finding; nor did it indicate that the District Court improperly equated education with skill.

36. Dan L. Burk and Mark A. Lemley, "Policy Levers in Patent Law," *Virginia Law Review* 89 (2003): 1575–1696, claim that the Federal Circuit is likely to play a corrective role here by being prone to find software patents obvious, but "those that it does approve will be entitled to broad protection." This is consistent with the recommendations I make in Chapter 8.

37. We are starting to see hints that the law may move in this direction, though the waters are far from clear. For example, the Supreme Court hinted that "if a person of ordinary skill *can implement* a predictable variation, § 103 likely bars its patentability." *KSR Int'l, Co. v. Teleflex, Inc.*, 127 S.Ct. 1727 (2007; emphasis added). This would seem to shift the focus away from what a person having ordinary skill in the art *would have thought of*, and toward what such a person is *capable of producing*. Computer Science Professor Jordan Pollack at Brandeis suggested a similar test under which an invention is only nonobvious if it "cannot be discovered by computer search process." Jordan Pollack, "Seven Questions for the Age of Robots," Yale Bioethics Seminar, January 2004, http://www.jordanpollack.com/sevenlaws.htm (accessed May 30, 2008). Although my recommendation is not the same as Pollack's, mine might be an obvious variation on his.

38. The Supreme Court has recently reaffirmed the desirability of disclosure because it forms the basis for further innovation (see note 23). The corollary, however, is that "these advances, once part of our shared knowledge, define a new threshold from which innovation starts once more. And as progress beginning from higher levels of achievement is expected in the normal course, the results of ordinary innovation are not the subject of exclusive rights under the patent laws. Were it otherwise patents might stifle, rather than promote, the progress of useful arts" (ibid.).

39. The test also flexibly takes into account the state of inventive technology extant at the time of invention. See Brief of Economists and Legal Historians as *Amici Curiae* in Support of Petitioner, at 13, *KSR Inter. Co. v. Teleflex, Inc.*, 127 S. Ct. 1727 (2007), arguing that "exogenous developments in technology" should be taken into account when ascertaining the current level of ordinary skill.

40. *KSR*, 127 S. Ct. at 16 ("The question is not whether the combination was obvious to the patentee but whether the combination was obvious to a person with ordinary skill in the art").

Chapter 8

1. *Amgen v. Chugai*, 927 F.2d 1200, 1206 (Fed. Cir. 1991).

2. *In re Hayes Microcomputer Prods. Inc. Patent Litigation*, 982 F.2d 1527, 1533–1536 (Fed. Cir. 1992).

3. See, e.g., Theresa Howard, "Gillette Ups Ante on Whisker Removal," *USA*

Today, January 15, 2004, http://www.usatoday.com/money/industries/retail/2004-01-15-electric-mach3_x.htm (accessed May 29, 2008).

4. See James M. Kilts (pseudonym), "Fuck Everything, We're Doing Five Blades," *Onion*, February 18, 2004, http://www.theonion.com/content/node/33930 (accessed May 29, 2008).

5. *O'Reilly* v. *Morse*, 56 U.S. 62, 112 (1854).

6. In practice, it might be better to require the inventor to demonstrate that her wish can be used to generate more than one design but fewer than all *possible* designs, because the latter might be impossible to prove even in theory.

7. "The scope of [a patent claim] must bear a reasonable correlation to the scope of enablement provided by the specification to persons of ordinary skill in the art." *In re Fisher*, 427 F.2d 833, 839 (C.C.P.A. 1970).

8. The Federal Circuit considers the level of disclosure necessary to support a claim to depend on the state of the art. See *Capon* v. *Eshhar*, 418 F.3d 1349, 1357 (Fed. Cir. 2005) ("The descriptive text needed to meet these requirements varies with the nature and scope of the invention at issue, and with the *scientific and technologic knowledge already in existence*"; emphasis added).

9. See Jessica Silbey, "The Mythical Beginnings of Intellectual Property," *George Mason Law Review* 15 (2008): 319–379.

10. See Campbell-Kelly, 107–108, for a discussion of the relative advantages and disadvantages of patent, copyright, and trade secret as they played out in the software industry in the 1960s.

11. See, e.g., Samuelson et al., 2327.

12. See Plotkin, "A History of Software Patents."

13. To take this line of reasoning to its logical conclusion, we can imagine a wish actually *being* an end product. Imagine that artificial invention technology has advanced to the point where it can grant your wish for an antilock braking system in a microsecond. Now you can put the wish on a disk (or on paper) inside a genie-equipped computer housed under the engine of a car. When you subsequently drive the car and slam on the brakes, your wish is transformed into antilock braking software, which then engages the car's brakes. From your perspective as the driver, there is no difference between this scenario and one in which the car came equipped with a hardwired antilock braking system from the start. The wish and the wish-come-true have converged.

14. See Samuelson et al., 2314.

15. Software was not seen as having any independent economic value through the 1950s (Campbell-Kelly, 54). It was not until IBM decided to "unbundle" pricing of its software from its hardware that an independent software industry emerged (ibid., 109–119).

16. "The object of the patent law in requiring the patentee [to distinctly claim his invention] is not only to secure to him all to which he is entitled, but to apprise the public of what is still open to them." *McClain* v. *Ortmayer*, 141 U.S. 419, 424 (1891).

17. Professor Michael Meurer and Lecturer James Bessen of Boston University School of Law have characterized vague patents as creating "fuzzy boundaries" that

impede such patents from properly serving their role as property. James Bessen and Michael Meurer, *Patent Failure: How Judges, Bureaucrats, and Lawyers Put Innovators at Risk* (Princeton: Princeton University Press, 2008), 46–72. Clear patent boundaries promote innovation for the same reason good fences make good neighbors (ibid., 20).

18. Bessen and Meurer, 20.

19. See Bessen and Meurer, 187–214.

20. See Plotkin, "A History of Software Patents"; Bessen and Meurer, 187–214.

21. Bessen and Meurer, 189.

22. William Neukom, "Microsoft Statement on Software Patents," January 26–27, 1994, http://www.jamesshuggins.com/h/tek1/software_patent_microsoft. htm (accessed June 2, 2008); "Oracle Corporation Patent Policy," 1994, http://www. jamesshuggins.com/h/tek1/software_patent_oracle.htm (accessed June 2, 2008); Warren, "Autodesk Statement on Software Patents."

23. 409 U.S. 63 (1972).

24. John Perry Barlow, "A Declaration of the Independence of Cyberspace," http://homes.eff.org/barlow/Declaration-Final.html (accessed May 29, 2008).

25. Barlow, "The Economy of Ideas."

26. James Gleick, "Patently Absurd," *New York Times Magazine*, March 12, 2000, 44.

27. For a general overview of the Compton's Multimedia patent, see Terri Suzette Hughes, "Patent Reexamination and the PTO: Compton's Patent Invalidated at the Commissioner's Request," *John Marshall Journal of Computer and Information Law* 14 (1996): 379-408; see also Simson L. Garfinkel, "Patently Absurd," *Wired*, July 1994, http://www.wired.com/wired/archive/2.07/patents_pr.html (accessed June 2, 2008); Seth Shulman, *Owning the Future* (New York: Houghton Mifflin, 1989), 59–62.

28. See, e.g., Tim Richardson, "BT Launches US Hyperlinks Legal Action," *Register*, December 14, 2000, http://www.theregister.co.uk/2000/12/14/bt_launches_us_ hyperlinks_legal/ (accessed May 28, 2008).

29. The directive was eventually defeated in 2005. Lucy Sherriff, "EU Parliament Bins Software Patent Bill," *Register*, July 6, 2005, http://www.theregister. co.uk/2005/07/06/eu_bins_swpat/ (accessed May 31, 2008).

30. See, e.g., *In re Slayter*, 276 F.2d 408 (C.C.P.A. 1960).

31. The Patent Office performs a similar analysis in evaluating the patentability of specific chemicals in light of prior claims to the genus or class to which they pertain. See, e.g., *In re Baird*, 16 F.3d 380 (Fed. Cir. 1994).

32. See *Graver Tank & Mfg. Co.* v. *Linde Air Prods. Co.*, 339 U.S. 605 (1950); *Eibel Process Co.* v. *Minnesota & Ontario Paper Co.*, 261 U.S. 45 (1923).

33. U.S. Patent No. 4,405,829 (issued September 20, 1983). The RSA patent is an example of a broad software patent that has withstood court challenges to its validity. See, e.g., *Schlafly* v. *Caro-Kann Corp.*, 1998 U.S. App. LEXIS 8250 (Fed. Cir. 1998).

34. It has since been learned that the algorithm was independently invented by a British government mathematician, Clifford Cocks, in 1973. Cocks, however, considered his creation to be classified and therefore kept it secret. Cocks's work was not

made public until 1997. Rivest, Shamir, and Adleman therefore invented the RSA algorithm independently of Cocks. Peter Wayner, "British Document Outlines Early Encryption Discovery," *New York Times*, December 24, 1997, http://www.nytimes.com/library/cyber/week/122497encrypt.html (accessed October 13, 2008).

35. See Don Marti, "Good-Bye Bandits, Hello Security," *Linux Journal*, October 1, 2000, http://www.linuxjournal.com/article/4254 (accessed October 13, 2008).

36. Under U.S. patent law, a description of an invention in a patent must enable a person having ordinary skill in the relevant field (art) to put the invention into practice without engaging in "undue experimentation." *W.L. Gore & Associates, Inc.* v. *Garlock*, 721 F.2d 1540, 1557 (Fed. Cir. 1983).

37. See Robert Plotkin, "A History of Software Patents," in *The History of Information Security*, eds. Karl de Leeuw and Jan Bergstra (Amsterdam: Elsevier, 2007), 141–164.

38. 151 Eng. Rep. 1266 (1841).

39. 55 U.S. 156 (1852).

40. 306 U.S. 86 (1938).

41. See, e.g., *In re Schrader*, 22 F.2d 290 (Fed. Cir. 1994).

42. Such patent claims are called "product-by-process" claims, because they describe a product in terms of the process used to make it. See, e.g., *Scripps Clinic & Research Foundation* v. *Genentech, Inc.* 927 F.2d 1565 (Fed. Cir. 1991); *Atlantic Thermoplastics Co., Inc.* v. *Faytex Corp.*, 970 F.2d 834 (Fed. Cir. 1992).

43. The existence of situations in which inventors could not describe their inventions in structural terms was the principal justification for permitting product-by-process claims. *Atlantic Thermoplastics Co., Inc.* v. *Faytex Corp.*, 970 F.2d 834, 843 (Fed. Cir. 1992). Eventually, however, the Patent Office and courts permitted applicants to draft claims in product-by-process format without demonstrating that the product could not be claimed structurally. See, e.g., *In re Steppan*, 394 F.2d 1013, 1019 (C.C.P.A. 1967). For more information on product-by-process claims, see Chisum, § 8.05.

44. See 35 U.S.C. § 112 ¶6.

45. See Michael J. Meurer, "Business Method Patents and Patent Floods," *Washington University Journal of Law and Policy* 8 (2002): 309–339.

46. Ina Fried, "Gates Wants Patent Power," CNET News, July 29, 2004, http://news.cnet.com/Gates-wants-patent-power/2100-1014_3-5288722.html (accessed October 13, 2008).

47. Josh Lerner, "Where Does State Street Lead? A First Look at Finance Patents, 1971–2000," *Journal of Finance* 57 (April 2002): 901–930.

48. U.S. Patent No. 6,847,851 (issued January 25, 2005).

49. U.S. Patent No. 7,117,186 (issued October 3, 2006).

50. U.S. Patent No. 7,047,169 (issued May 16, 2006).

51. "The Hierarchical Bayesian Optimization Algorithm," http://www.illigal.uiuc.edu/hboa/ (accessed October 13, 2008).

52. See James Bessen, "Patent Thickets: Strategic Patenting of Complex Technologies" (March 2003), http://ssrn.com/abstract=327760 (accessed June 2, 2008).

Chapter 9

1. See Annabelle Gawer and Michael A. Cusumano, *Platform Leadership: How Intel, Microsoft, and Cisco Drive Industry Innovation* (Boston: Harvard Business School Press, 2002), 2–3 (defining a platform as "an evolving system made of interdependent pieces that can each be innovated upon").

2. See Gawer and Cusumano, 15–76, 131–187, for an overview of the strategies employed by Intel and Microsoft to achieve "platform leadership." Sometimes people refer to the operating system itself (e.g., Microsoft Windows) as a "platform." See, e.g., Steven Weber, *The Success of Open Source* (Cambridge, MA: Harvard University Press, 2004), 9.

3. See Lawrence Lessig, *The Future of Ideas* (New York: Vintage, 2001), 23 (describing the Internet as an "innovation commons").

4. See, e.g., ibid.

5. For an overview of arguments for and against the claim that open platforms promote more innovation than closed platforms do, see Kevin Boudreau, "Does Opening a Platform Generate More Innovation? An Empirical Study," MIT Sloan Research Paper No. 4611 (2007). Boudreau concludes, from an empirical study of the handheld computer industry, that exercising some control over and closure of a platform can lead to more innovation than leaving the platform entirely open.

6. Sara Boettinger and Dan L. Burk use the term *open* in a similar sense when they point out that in the phrase "open source" the word "refers to a certain philosophy of access, improvement, and production." They note that a patented invention may not be open in this sense because the invention is not available or accessible to the public, even though the patent document makes information about the invention available to the public. Sarah Boettinger and Dan L. Burk, "Open Source Patenting," *Journal of International Biotechnology Law* 1 (2004): 225; see also Andrew Hessel, "Open Source Biology," in *Open Sources 2.0*, ed. Chris DiBona et al. (Sebastopol, CA: O'Reilly, 2006), 281–296. My usage of *open* is also consistent with the basic common denominator among open source licenses: "the software's source code is always freely available and users can modify it without restriction." See Stephen R. Walli, "Under the Hood: Open Source and Open Standards Business Models in Context," in ibid., 126.

7. See, e.g., Lawrence Lessig, *Code and Other Laws of Cyberspace* (New York: Basic Books, 1999), 102–108.

8. The World Wide Web Consortium (W3C) maintains detailed technical documentation regarding the HTTP protocol. See W3C, "HTTP: Hypertext Transfer Protocol," W3C, http://www.w3.org/Protocols/ (accessed May 29, 2008).

9. Free Software Foundation, "The Free Software Definition," Free Software Foundation, http://www.fsf.org/licensing/essays/free-sw.html (accessed May 29, 2008).

10. Eric Raymond, Larry Augustin, and John Hall coined the term *open source* in February 1998 (Weber, 114).

11. Steven Weber claims that open source software is "an experiment in social organization around a distinctive notion of property" in which a property right is a

right to *distribute* the object of the property right, rather than the traditional right to *exclude others* from using the object of the property right (Weber, 1).

12. Weber proposes four "organizational principles for distributed innovation" suggested by the successes of open source software, which overlap somewhat with the list in the text, such as "empower people to experiment" and "enable bits of information to find each other" (Weber, 235).

13. 35 U.S.C. § 102(b); *Feist* v. *Rural Tel. Serv. Co.*, 499 U.S. 340 (1991).

14. *Diamond* v. *Diehr*, 185.

15. See, e.g., *Brenner* v. *Manson*, 383 U.S. 519 (1966). See also Julian David Forman, "A Timing Perspective on the Utility Requirement in Biotechnology Patent Applications," *Albany Law Journal of Science and Technology* 12 (2002): 647–682; Dan L. Burk and Mark A. Lemley, "Policy Levers in Patent Law," *Virginia Law Review* 89 (2003): 1644–1646.

16. See *Diamond* v. *Chakrabarty*, 447 U.S. 303, 309 (1980).

17. You might find it surprising to learn that a work in the public domain, such as a product that has never been patented and can no longer be patented by anyone, is actually not as open as a true open source product. The reason is that someone can still modify the public domain product and then patent the modified product, thereby making the modified product closed. Truly open products cannot "go closed" in this way.

18. For some examples, see the toothbrush and controller described in the Introduction.

19. Louis Kronenberger, *Viking Book of Aphorisms*, ed. W. H. Auden (Viking Adult, 1962), 111.

20. In the law of real property, this is called an "affirmative easement." *Restatement (Third) of Property (Servitudes)* § 1.2 (1998).

21. For example, IBM recently made the technology in 500 of its patents available for use by open source software developers. Robert McMillan, "Developers Voice Mixed Reactions to IBM Patent Policy," *IDG News Service*, January 12, 2005, http://www.infoworld.com/article/05/01/12/HNpatentreaction_1.html (accessed May 29, 2008). Similarly, Sun Microsystems recently took similar steps in connection with a large number of its patents. Paul Krill, "Sun Introduces OpenSolaris, Releases 1,670 Patents," *InfoWorld*, January 25, 2005 (accessed May 29, 2008).

22. The opposite of open is closed, not proprietary. Similarly, the opposite of proprietary is public domain or nonproprietary. See Doc Searls, "Making a New World," in Dibona et al., 247 (quoting Craig Burton).

23. The Semiconductor Chip Protection Act of 1984, 17 U.S.C. §§ 901–914 (2007).

24. See Leon Radomsky, "Sixteen Years After the Passage of the U.S. Semiconductor Chip Protection Act: Is International Protection Working?" *Berkeley Technology Law Journal* 15 (2000): 1049–1094.

25. Open platforms do not always spur more innovation than closed platforms. Just take Apple, "arguably the most successful company in Silicon Valley these days, yet its practices are at odds with a prevailing assumption in most of techland that

open-platform collaboration and iterative group work are what drive creativity and growth." Robert Safian, "Forbidden Fruit," *Fast Company*, December 2007, 18.

26. See generally *State Street Bank & Trust Co. v. Signature Financial Group, Inc.*, 149 F.3d 1368 (Fed. Cir. 1998).

27. 56 U.S. 62, 113 (1854).

28. 409 U.S. 63, 67 (1972).

29. See Chris DiBona, "Open Source and Proprietary Software Development," in DiBona et al., 34n9.

30. One of the most popular open source licenses is the GNU General Public License (GPL), which was originally developed by Richard Stallman and released in January 1989. The most recent revision, GPL v. 3, was released on June 29, 2007. Free Software Foundation, "FSF Releases the GNU General Public License, Version 3," Free Software Foundation, June 29, 2007, http://www.fsf.org/news/gplv3_launched (accessed May 29, 2008).

31. For an overview of "copyleft" as described by the Free Software Foundation, see "Licenses," Free Software Foundation, http://www.gnu.org/licenses/ (accessed May 29, 2008).

32. Donald S. Chisum, *Patents, a Treatise on the Law of Patentability, Validity and Infringement* (New York: Matthew Bender, 2000), § 4.02 (citing *In re Swartz*, 232 F.3d 862 (Fed. Cir. 2000)).

33. U.S. Patent and Trademark Office, "Utility Examination Guidelines," 66 FR 1092 (2001).

34. See U.S. Patent and Trademark Office, *Manual of Patent Examining Procedure*, 8th ed. (Alexandria, VA: U.S. Patent and Trademark Office, 2007), § 2107. The credibility "requirement excludes 'throw-away,' 'insubstantial,' or 'nonspecific' utilities, such as the use of a complex invention as landfill, as a way of satisfying the utility requirement of 35 U.S.C. 101."

35. *In re Fisher*, Appeal 04–1465, 14, September 7, 2005.

36. See, e.g., Andy Oram, "Patent Pools Offer Open Source a New Incentive—and a New Source of Power," O'Reilly ONLamp.com, December 8, 2005, http://www.oreillynet.com/onlamp/blog/2005/12/patent_pools_offer_open_source.html (accessed May 28, 2008); OASIS, "Open Source Development Labs (OSDL) Announces Patent Commons Project," Cover Pages, August 10, 2005, http://xml.coverpages.org/ni2005-08-10-a.html (accessed May 28, 2008); "Pundits Dispute Effectiveness of Open-Source Patent Pools," LinuxElectrons, November 19, 2005, http://www .linuxelectrons.com/news/general/pundits-dispute-effectiveness-open-source-patent-pools (accessed May 28, 2008).

Chapter 10

1. 35 U.S.C. § 112 ¶1.

2. The more technical way of saying this is that you should be able to build the frame without engaging in undue experimentation. The idea is that building the

frame, using the patent as a blueprint, should not require *you* to engage in any inventive effort. If you do, then the patent owner has not provided sufficient information to satisfy the enablement requirement. See *Kewanee Oil Co. v. Bicron Corp.*, 416 U.S. 470, 480–484 (1974).

3. The U.S. Patent Office did require models to be submitted until the late 19th century, but stopped the practice after fires destroyed most of the models and the increasing number of patent applications made the practice unwieldy.

4. If the artificial genie is not readily available to the public, the Patent Office could refuse to issue the patent on the grounds that the inventor has not furnished an "enabling" disclosure, as required by 35 U.S.C. § 112 ¶1.

5. See, e.g., *Fiers* v. *Revel*, 984 F.2d 1164, 1170 (Fed. Cir. 1993); *The Regents of the Univ. of Cal.* v. *Eli Lilly & Co.*, 119 F.3d 1559, 1568–1569 (Fed. Cir. 1997).

6. Such descriptions of actual experimental results would not only put the public on notice of the full scope of the claim but also demonstrate that the inventor has satisfied the enablement requirement for the full scope of the claim. Because real-world artificial genies have limitations and always will, the mere fact that a wish can be used to produce *one* product (e.g., a car frame) using a state-of-the-art artificial genie does not imply that the same wish can be used to produce *all possible* car frames, or even a range of car frames, falling within the scope of the wish.

A broad wish claim encompassing multiple physical variations might be considered to be enabled only if the wish works every time. Consider the patent in *In re Goodman*, 11 F.3d 1046 (Fed. Cir. 1993), in which the patentee broadly claimed a genetic expression technique. He described only a single working example in tobacco plants, but the patent claim purported "to cover any desired mammalian peptide produced in any plant cell" (ibid., 1049). "This single example, however, does not enable a biotechnician of ordinary skill to produce any type of mammalian protein in any type of plant cell."

7. In the United States, the requirement that inventors disclose relevant prior art to the Patent Office might already require inventors to disclose to the Office any well-known or obvious wishes that could be used to produce the artificial inventions for which they seek patent protection.

8. See, e.g., James Gleick, "Patently Absurd," *New York Times*, March 12, 2000, http://www.nytimes.com/library/magazine/home/20000312mag-patents.html (accessed June 2, 2008); Jim Warren, Prepared Testimony and Statement for the Record of Jim Warren Before the Patent and Trademark Office, January 26–27, 1994, http://www.uspto.gov/web/offices/com/hearings/software/sanjose/sj_warren.html (accessed May 28, 2008); Mark Aaron Paley, "A Model Software Petite Patent Act," *Computer and High Technology Law Journal* 12 (1996): 306; Randall Davis et al., "A New View of Intellectual Property and Software," *Communications of the ACM* 39, no. 3 (March 1996): 28.

9. See, e.g., Julie E. Cohen and Mark A. Lemley, "Patent Scope and Innovation in the Software Industry," *California Law Review* 89 (2001): 1; Paley, 343–346.

10. This would address Ben Klemens's concern in *Math You Can't Use* that

mathematicians and programmers can be liable for infringing a patent even for engaging in work performed without knowledge of the patent; see Klemens, 158.

11. Paley, 334–335.

12. See, e.g., "Oracle Corporation Patent Policy," Patent & Trademark Office Software Patent Hearings (Jan. 26–27, 1994), http://www.jamesshuggins.com/h/tek1/software_patent_oracle.htm (accessed May 28, 2008); Richard H. Stern, "An Attempt to Rationalize Floppy Disk Claims," *John Marshall Journal of Computer and Information Law* 17 (1998): 183.

13. See, e.g., Warren; Simson L. Garfinkel, "Patently Absurd," *Wired*, July 1994, http://www.wired.com/wired/archive/2.07/patents.html (accessed May 28, 2008); Paley, 306.

14. For examples of these and other reforms, see Tim O'Reilly, "The Internet Patent Land Grab," *Communications of the ACM* 43, no. 6 (June 2000): 29–31.

15. See, e.g., O'Reilly Media, Inc., "Internet Society Panel on Business Method Patents," The O'Reilly Network, Oct. 23, 2000, http://www.oreillynet.com/pub/a/policy/2000/10/23/isoc.html (accessed May 29, 2008); U.S. Patent and Trademark Office, "Business Methods Patent Initiative: An Action Plan," September 14, 2003, http://www.uspto.gov/web/offices/com/sol/actionplan.html (accessed May 29, 2008). Furthermore, the American Inventors Protection Act, Pub. L. 106–113, modified 35 U.S.C. § 154(d) of the U.S. Patent Act to require that most patent applications be published 18 months after they are filed, providing the public with earlier notice of the subject matter potentially covered by a patent and making it easier for the public to challenge the patentability of an invention claimed in a patent application. See 37 CFR §§ 1.902–1.997.

16. For more information on the USPTO's classification system, see U.S. Patent and Trademark Office, "Handbook of Classification," March 2005, http://www.uspto.gov/web/offices/opc/documents/handbook.pdf (accessed May 29, 2008).

17. The fair use defense is codified at 17 U.S.C. § 107. The independent creation defense arises out of the originality requirement, which the U.S. Supreme Court has declared is a constitutional requirement for copyright protection. *Feist Publ'ns, Inc.* v. *Rural Tel. Serv. Co., Inc.*, 499 U.S. 340 (1991).

18. Currently, copyright protection for newly published works "endures for a term consisting of the life of the author and 70 years after the author's death." 17 U.S.C. § 302(a). In contrast, patents extend for a term of "20 years from the date on which the application for the patent was filed in the United States." 35 U.S.C. § 154(b).

19. See, e.g., Pamela Samuelson et al., "A Manifesto Concerning the Legal Protection of Computer Programs," *Columbia Law Review* 94 (1994): 2308–2431.

20. Jeffrey Ullman, "Ordinary Skill in the Art," November 16, 2000 (minor updates August 30, 2001), Stanford University Infolab, http://infolab.stanford.edu/ullman/pub/focs00.html (accessed May 29, 2008).

21. See Peer to Patent: Community Patent Review Project, http://www.peertopatent.org/ (accessed May 29, 2008).

22. For a thorough discussion, see Dan L. Burk and Mark A. Lemley, "Is Patent Law Technology-Specific?" *Berkeley Technology Law Journal* 17 (2002): 1155–1206;

Dan L. Burk and Mark A. Lemley, "Policy Levers in Patent Law," *Virginia Law Review* (2003): 1575–1696.

23. See *In re Deuel*, 51 F.3d 1552, 1559 (Fed. Cir. 1995).

24. See David E. Goldberg, *The Design of Innovation* (Boston: Kluwer Academic, 2002), 227–228.

25. See Dror G. Feitelson, "Experimental Computer Science," *Communications of the ACM* 50, no. 11 (November 2007): 24–26.

26. For overviews of how biology and computer science are increasingly influencing one another in both directions, see, e.g., Peter J. Bentley, *Digital Biology: How Nature Is Transforming Our Technology and Our Lives* (New York: Simon & Schuster, 2001); Janine M. Benyus, *Biomimicry: Innovation Inspired by Nature* (New York: Perennial 2002); Nancy Forbes, *Imitation of Life: How Biology Is Inspiring Computing* (Cambridge, MA: MIT Press, 2004); Moshe Sipper, *Machine Nature: The Coming Age of Bio-Inspired Computing* (New York: McGraw-Hill, 2002).

27. Jon Cohen, "Sequencing in a Flash," *MIT Technology Review*, May/June 2007, 72–77.

28. See Thomas Goetz, "The Age of the Genome," *Wired*, November 17, 2007, 256–265; Peter Aldhous, "Your Own Book of Life," *New Scientist*, September 8, 2007, 8–11.

29. Chris Taylor, "How Biotech Is Driving Computing," *Business 2.0*, August 18, 2006, http://money.cnn.com/2006/08/18/technology/futureboy0818.biz2/index.htm (accessed June 2, 2008).

30. U.S. Patent No. Re. 32,580 (reissued Jan. 19, 1988).

31. More precisely, the parties stipulated that Microsoft infringed AT&T's patent for purposes of the appeal. *AT&T v. Microsoft*, 414 F.3d 1366, 1368 (Fed. Cir. 2005).

32. Ibid.

33. 35 U.S.C. § 271(f).

34. Ibid., 1368–1369.

35. Ibid., 1370.

36. Jordan Pollack, interview by author, July 13, 2005.

37. "Specialization Is for Insects." Robert A. Heinlein, *Time Enough for Love* (New York: Penguin Putnam, 1973), 248.

38. See Donald E. Knuth, "Computer Programming as an Art," *Communications of the ACM* 17, no. 12 (December 1974): 667–673, for a discussion of the history of the terms *art* and *science*. In particular, Knuth points out that although art and science once had the same or very similar meanings, they gradually took on independent meanings over time.

39. See generally Dan L. Burk, "Patenting Speech," *Texas Law Review* 79 (2000): 99–135.

40. *Ex Parte* Lundgren, Appeal No. 2003–2088 (2004).

41. Steven G. Kunin, "Patent Eligibility 35 U.S.C. § 101 for Non-machine Implemented Processes," slide 42 (presented at "USPTO Day" at the U.S. Patent and Trademark Office in December 2004).

42. Ibid., slide 50.

43. The Court of Appeals recently clarified that "systems that depend for their operation on human intelligence *alone*" (emphasis added) are not patentable. *In re Comiskey*, 499 F.3d 1365 (Fed. Cir. 2007). Such a decision, however, does not expressly foreclose obtaining patents on systems and processes that involve even minimal manipulation of physical material.

44. See Richard Susskind, *The Future of Law* (New York: Oxford University Press, 1998).

45. Pamela Samuelson et al., "A Manifesto Concerning the Legal Protection of Computer Programs," *Columbia Law Review* 94 (1994): 2308–2431.

46. Samuelson, 2408.

47. Martin Campbell-Kelly argues that "the concept of a 'software industry' may one day be too fuzzy to be meaningful," once programming is adopted as a means of creating a sufficiently broad range of products. Campbell-Kelly, 301.

Chapter 11

1. For an overview of some of the potential harmful consequences of outsourcing and automation for engineers, see Stephen H. Unger, "Making Computer Professionals and Other Engineers Low-Priced Commodities," *IEEE Technology and Society*, Summer 2004, 36–40.

2. See Napoleon Hill, *Think and Grow Rich* (New York: Fawcett Crest, 1960), 28–29, for an entertaining, if not necessarily historically accurate, history of the V-8's creation.

3. It is often claimed that writing software is easier than designing hardware. For example, Richard Stallman and Simson Garfinkel claim that "software systems are much easier to design than hardware systems of the same number of components." The League for Programming Freedom, "Against Software Patents," *Communications of the ACM* 35, no. 1 (January 1992): 17–22, 121. Comparing only software and hardware *of the same number of components*, however, is misleading precisely because computers automate the process of transforming a small number of high-level software instructions into a large number of low-level components. If we are concerned about whether the daily work of programmers is any easier than the daily work of hardware engineers, then we should instead compare the difficulties of the problems that each group attempts to tackle in its own field.

4. Some programmers argue that computer programming is inherently easy. For example, Ben Klemens claims that "there is no magic or genius to the process of programming, just small components built upon larger structures, and then still larger structures built upon those." Klemens, *Math You Can't Use* (Harrisonburg, VA: Brookings Institution Press, 2005), 24. This argument is flawed; mechanical inventing can involve genius even though most, if not all, mechanical inventing involves combining small components into larger structures and those into yet larger structures.

5. According to Christof Teuscher, the fitness function is the most important ele-

ment in most uses of genetic algorithms: "It is a key element because in that fitness function you can build in knowledge about the solution you want." Since the fitness function guides the simulated evolutionary process, you can use it to exclude portions of the search space and thereby focus and speed up the search. Christof Teuscher, interview by author, June 13, 2005. Similarly, Peter Bentley told me, "Most of our cleverest automatic inventions have really been helped quite a lot. It's actually very hard to set up an evolutionary algorithm to create something brand new without giving it any information at all. It's my preferred way of doing it, but it's very hard to actually get a complex result out without cheating by giving it part of the output first." Peter Bentley, interview by author, July 22, 2005.

6. Gregory Hornby, one of the NASA engineers responsible for the antenna we saw in the Introduction, spends much of his time creating better simulators and software for evaluating the results of the simulations. Gregory Hornby, interview by author, June 28, 2005.

7. The core team at Matrix Advanced Solutions includes people with Ph.D.s in physics, math, and biology, with strong capabilities in algorithmic development and programming. Matrix A/S's Dr. Marcel Thuerk and Sion Balass believe that the most valuable skills to have in their field are in "applied sciences," which they define as falling between pure science and engineering. Marcel Thuerk and Sion Balass, interview by author, November 15, 2007; Sion Balass, interview by author, October 26, 2005.

8. For more examples of bio-inspired computing, see Taylor, "How Biotech Is Driving Computing."

9. See Thomas Hoffman, "What Tech Skills Are Hot for 2006?" *Computerworld*, December 7, 2005, http://www.computerworld.com/careertopics/careers/story/0,10801,107363,00.html (accessed on May 28, 2008).

10. For more on the value of such skills, see generally Dan Pink, *A Whole New Mind* (New York: Riverhead Books, 2006). For several examples of how pre-artificial invention computer technology has been used to solve problems in fields ranging from pure mathematics to medicine to game playing to gene sequencing, see S. Sadagopan, "Computer Science Growing into a Basic Science," *Financial Express*, January 20, 2006, http://www.financialexpress.com/news/Computer-science-growing-into-a-basic-science/63264/2 (accessed May 28, 2008). David E. Goldberg refers to this entrepreneur of tomorrow as the "entrepreneurial engineer." See Goldberg, *The Entrepreneurial Engineer* (Hoboken, NJ: Wiley, 2006).

11. Pink, p. 126.

12. Jeff Walker, interview by author, March 9, 2006.

13. David Fogel, interview by author, September 20, 2007.

14. Ibid.

15. Steve Lohr, "A Techie, Absolutely, and More," *New York Times*, August 23, 2005, http://www.nytimes.com/2005/08/23/technology/23geeks.html?pagewanted=all (accessed June 2, 2008).

16. See Stephen H. Unger, "Making Computer Professionals and Other Engineers Low-Priced Commodities," *IEEE Technology and Society*, Summer 2004, 36–40.

17. David Fogel, interview by author, September 20, 2007.

18. Moshe Sipper, *Machine Nature: The Coming Age of Bio-Inspired Computing* (New York: McGraw-Hill, 2002), 187.

19. The Office of Charles and Ray Eames, *A Computer Perspective* (Cambridge, MA: Harvard University Press, 1990), 101.

20. Essinger, 40.

21. See, e.g., "Reverse Outsourcing for Indian Call Centres," *Outsourcing Times*, June 14, 2005, http://www.blogsource.org/2005/06/reverse_outsour_1.html (accessed May 28, 2008).

22. Robert L. Glass, "The Plot to Deskill Software Engineering," *Communications of the ACM* 48, no. 11 (November 2005): 21–24. See also Simon Head, *The New Ruthless Economy: Work and Power in the Digital Age* (New York: Oxford University Press, 2003), 10 (describing how information technology can enhance human skills ("skill complementarity"), replace human skills ("skill substitution"), or reduce the level of human skill exercised to perform a given function ("skill debilitation").

23. See, e.g., Head, 72–75.

24. Quoted in Paul Armer, "Attitudes Toward Intelligent Machines," in *Computers and Thought*, ed. Edward A. Feigenbaum and Julian Feldman (Menlo Park, CA: MIT Press, 1995), 389–405. Automated computer technology can also give employers and others greater ability to monitor and control creative workers in much the same way that Frederick Taylor's scientific management was designed to monitor and control the manual workforce of the early 20th century. See Head.

25. Eames, 34.

26. Eames, 34, quoting Dorr E. Felt.

27. Rheingold, 16.

28. David Talbot, "Luis von Ahn," *MIT Technology Review*, September/October 2007, 61.

29. Karl Sims, "Galápagos," http://www.genarts.com/galapagos/index.html (accessed May 28, 2008); Karl Sims, interview by author, May 19, 2005.

30. Adam L. Penenberg, "Man vs. Machine," *Fast Company*, September 2007, 97–102.

31. Frank Levy and Richard J. Murnane, *The New Division of Labor: How Computers Are Creating the Next Job Market* (Princeton: Princeton University Press, 2004), 4.

32. See David E. Goldberg, *The Design of Innovation*, 228.

Chapter 12

1. "Althea Technologies, Inc. and Natural Selection, Inc. Introduce Time Saving Assay for Predicting Carcinogenicity Using PCR-Based Gene Expression Profiling," November 8, 2005, http://www.natural-selection.com/Press/2005/pr11082005.htm (accessed May 28, 2008).

2. "2005 ComputerWorld Honors Program," *ComputerWorld*, http://www.cwhonors.org/search/his_4a_detail.asp?id=511 (accessed May 28, 2008); "Configur-

ing a 500 Percent ROI for Dell," i2 Technologies, Inc., http://www.i2.com/customers/ success_stories/css.cfm?caid=73&cpageL1=hightech (accessed May 28, 2008); "Supply Chain Planning," i2 Technologies, Inc., http://www.i2.com/customers/ success_stories/css.cfm?caid=131&cpageL1=retail (accessed May 28, 2008); "Slashing Inventory at Whirlpool," i2 Technologies, Inc., http://www.i2.com/customers/ success_stories/css.cfm?caid=68&cpageL1=consumer_industries (accessed May 28, 2008).

3. Thomas Baeck, "Solving Multidisciplinary Optimization Problems in Product Engineering," *Electronic Products and Technology*, September 2006, http://www .nutechsolutions.com/pdf/Electronic%20Products%20and%20Technology.pdf (accessed May 28, 2008).

4. "Interview with Atomasoft Reporter Sander Olson," Imagination Engines, Inc., http://www.imagination-engines.com/atomasoft.htm (accessed May 28, 2008).

5. Sandra Upson, "Computer Software That Writes Itself," *Newsweek*, December 26, 2005, http://www.newsweek.com/id/51529 (accessed June 2, 2008).

6. Hod Lipson, interview by author, July 7, 2005.

7. Stephen H. Unger, "Making Computer Professionals and Other Engineers Low-Priced Commodities," *IEEE Technology and Society*, Summer 2004, 36–40.

8. Thomas W. Malone, *The Future of Work* (Boston: Harvard Business School Press, 2004), ix.

9. Ibid., 4. See also Daniel H. Pink, *Free Agent Nation: How America's New Independent Workers Are Transforming the Way We Live* (New York: Business Plus, 2002).

10. Pink, *Free Agent Nation*, 39.

11. For an overview of some of the effects of outsourcing on the IT industry in the United States in the 1990s, see Catherine L. Mann, "What Global Sourcing Means for U.S. IT Workers and for the U.S. Economy," *Communications of the ACM* 47, no. 7 (July 2004): 33–35.

12. See, e.g., David E. Gumpert, "An Atlas of Offshore Outsourcing," *Business Week*, February 18, 2004, http://www.businessweek.com/smallbiz/content/feb2004/ sb20040218_6502.htm (accessed June 2, 2008).

13. See generally Linus Torvalds, "The Linux Edge," in *Open Sources: Voices from the Open Source Revolution*, eds. Chris DiBona et al. (Sebastopol, CA: O'Reilly, 1999), 101–111; Malone, 41–43.

14. Jeff Howe, "The Rise of Crowdsourcing," *Wired* 14, no. 6 (June 2006): 176–183.

15. See generally Henry Chesbrough, *Open Innovation: The New Imperative for Creating and Profiting from Technology* (Boston: Harvard Business School Press, 2003).

16. Robert Wallace, "Early and Often: Harnessing the Consumer's Voice in Packaging Design," *Package Design*, November/December 2005, 48–51.

17. U.S. Patent No. 7,016,882 (issued March 21, 2006).

18. Richard Susskind, *The Future of Law* (New York: Oxford University Press, 1998).

19. U.S. Patent No. 6,607,389 (issued August 19, 2003).

20. U.S. Patent No. 7,076,439 (issued July 11, 2006).

21. U.S. Patent No. 6,574,645 (issued June 3, 2003).

22. According to Jeff Walker of TenFold, "what we've produced is sufficiently radical that it is hard for people to believe." Jeff Walker, interview by author, March 9, 2006. Similarly, Marcel Thuerk and Sion Balass of Matrix Advanced Solutions told me that people often initially think their technology is only science fiction. To convince people that their technology works, they have worked hard to build their credibility, as in successfully completing projects for the U.S. military. Marcel Thuerk and Sion Balass, interview by author, November 15, 2007; Sion Balass, interview by author, October 26, 2005.

23. Peter J. Bentley, *Digital Biology: How Nature Is Transforming Our Technology and Our Lives* (New York: Simon & Schuster, 2001), 56–57.

24. Moshe Sipper, *Machine Nature: The Coming Age of Bio-Inspired Computing* (New York: McGraw-Hill, 2002), 190–191.

Chapter 13

1. Although judges and legislators will also need to adapt to the Artificial Invention Age, I don't address those audiences separately in this chapter because the work they have cut out for them is implicit in the reinterpretations and changes to patent law that I describe in Part II.

2. Robert C. Faber, *Landis on Mechanics of Patent Claim Drafting*, 5th ed. (New York: Practicing Law Institute, 2004), 3–7.

3. U.S. Patent No. 6,049,811 (issued April 11, 2000).

4. See, e.g., *Fonar Corp.* v. *General Electric*, 107 F.3d 1543 (Fed. Cir. 1997).

5. Such proof would be useful not only for supporting the *validity* of the claim but also to support a broader *scope* of the claim. A patent on a wish that has been demonstrated to produce only a single car frame might be interpreted to cover only the particular frame. If, however, the inventor can demonstrate that the same wish can be used to automatically generate designs for a variety of new and useful car frames, then the patent may be interpreted more broadly to cover all of those designs. See Jeffrey L. Light, "Broadening the Scope of Biotechnology Patents by Disclosing a Scientific Theory," *Chicago-Kent Journal of Intellectual Property* 3 (2003): 87–113.

6. State contract law governs noncompetition agreements, with the precise contours varying from state to state.

7. Uniform Trade Secrets Act, § 1(4) (1985).

8. *Kewanee Oil Co.* v. *Bicron Corp.*, 416 U.S. 470, 494 (1974) (Marshall, J., concurring).

9. See James Bessen and Robert M. Hunt, "An Empirical Look at Software Patents," *Journal of Economics & Management Strategy* 16, no. 1 (2007): 157–189.

10. By definition, a trade secret must be the "subject of efforts that are reasonable under the circumstances to maintain its secrecy." Uniform Trade Secrets Act, § 1(4) (1985).

11. 35 U.S.C. § 154(a)(2).

12. Pamela Samuelson et al., "A Manifesto Concerning the Legal Protection of Computer Programs," *Columbia Law Review* 94 (1994), 2308–2431.

13. Technically, trade secret law prohibits "improper means" of competition while permitting "proper means." Uniform Trade Secrets Act, § 1 cmt. (1985); see also, *Kewanee Oil Co.*, 416 U.S., 476.

14. Julie E. Cohen and Mark A. Lemley, "Patent Scope and Innovation in the Software Industry," *California Law Review* 89 (2001): 1–57.

15. *Kewanee*, 416 U.S., 489–490.

16. For an overview of tradeoffs among patenting, publishing, and protecting information as a trade secret, see Iraj Daizadeh et al., "A General Approach for Determining When to Patent, Publish, or Protect Information as a Trade Secret," *Nature Biotechnology* 20 (October 2002): 1053–1054.

Chapter 14

1. For a detailed discussion of how users innovate, see Eric von Hippel, *Democratizing Innovation* (Cambridge, MA: MIT Press, 1990); Sonali K. Shah, "Open Beyond Software," in *Open Sources 2.0*, ed. Chris DiBona et al. (Sebastopol, CA: O'Reilly, 2006), 339–360.

2. Programming is also a kind of engineering to the extent that engineering is "the art or science of making practical application of the knowledge of pure sciences." Dick Hamlet and Joe Maybee, *The Engineering of Software: Technical Foundations for the Individual* (Boston: Addison-Wesley, 2001), 49–50.

3. Martin Campbell-Kelly, *From Airline Reservations to Sonic the Hedgehog: A History of the Software Industry* (Cambridge, MA: MIT Press, 2004), 212 (citation omitted).

4. Ibid., 213.

5. Ibid., 217.

6. Steven Weber, *The Success of Open Source* (Cambridge, MA: Harvard University Press, 2004), 25–26.

7. Campbell-Kelly, 223.

8. See, e.g., Ray Kurzweil, *The Singularity Is Near: When Humans Transcend Biology* (New York: Penguin, 2006), 56–84.

9. John Koza, email message to author, October 1, 2007.

10. John R. Koza, Sameer H. Al-Sakran, and Lee W. Jones, "Automated Reinvention of Six Patented Optical Lens Systems Using Genetic Programming," *Proceedings of the Genetic and Evolutionary Computation Conference, GECCO-2005* (New York: ACM Press, 2005).

11. Similarly, Eric Haseltine used a program called Zemax to design a lens for use in Disney theme parks, even though he did not know anything about optics or lens design. Haseltine, "Will Computers Replace Engineers?" *Discover*, February 1, 2003, http://discovermagazine.com/2003/feb/roundtable (accessed June 2, 2008).

12. See, e.g., Mike Nagle, "Games Consoles Reveal Their Hidden Power," *New Scientist*, February 16, 2008, 26–27.

13. Alvin Toffler coined the term *prosumer* in 1980 in *The Third Wave* (New York: Bantam, 1980).

14. Eric von Hippel, *Democratizing Innovation* (Cambridge, MA: MIT Press, 2005), 1–2. For further discussion of user innovation in sports products, see ibid., 27–30.

15. von Hippel, 19–22.

16. See, e.g., Paul Miller, "Sony Busts Down Mod-chip Retailer with $9 Mil. Lawsuit," *Engadget*, October 5, 2006, http://www.engadget.com/2006/10/05/sony-busts-down-mod-chip-retailer-with-9-mil-lawsuit (accessed May 28, 2008); Sam Varghese, "Playstation Mods Legal, Says High Court," *Age*, October 6, 2005, http://www.theage.com.au/news/breaking/playstation-mods-legal-says-high-court/2005/10/06/1128562920702.html (accessed May 28, 2008).

17. von Hippel, 108–109.

18. See von Hippel, 128–129, for examples.

19. See Jordan Ellenberg, "The Netflix Challenge," *Wired*, March 2008, 114–122.

20. For additional examples, see Mike Nagle, "Games Consoles Reveal Their Hidden Power" (sidebar entitled "Scientists Go out to Play"), *New Scientist*, February 16, 2008, 26–27.

21. Michael Patrick Gibson, "Democratizing Robot Design," *MIT Technology Review*, July/August 2007, 17.

22. Michael A. Prospero, "By the People, for the People," *Fast Company*, September 2007, 66–67.

23. Ibid., 66–67.

24. For background on the "e-lance" economy, in which individuals leverage electronic communications technology to work as freelancers, see Thomas W. Malone and Robert J. Laubacher, "The Dawn of the E-lance Economy," *Harvard Business Review* 76, no. 5 (September-October 1998): 144–152; Daniel Pink, *Free Agent Nation: The Future of Working for Yourself* (New York: Business Plus, 2002).

25. Campbell-Kelly, 50.

26. Neil Gershenfeld, *Fab: The Coming Revolution on Your Desktop—From Personal Computers to Personal Fabrication* (New York: Basic, 2005).

27. Gershenfeld, 3–4.

28. Gershenfeld, 39; Kurzweil, 227–228.

29. See Chris Anderson, *The Long Tail: Why the Future of Business Is Selling Less of More* (New York: Hyperion, 2006).

30. See generally, Christoph Berger and Frank Piller, "Customers as Co-Designers," *IEE Manufacturing Engineer*, August/September 2003, 42–45; Michael Patrick Gibson, "Automated Custom Manufacturing," *MIT Technology Review*, November 7, 2007, http://www.technologyreview.com/Biztech/19678/ (accessed June 2, 2008).

31. Video interview with Eric Bonabeau in Wade Rush, "The Art of the Possible," *MIT Technology Review*, September 1, 2006, http://www.technologyreview.com/Infotech/17397/ (accessed May 28, 2008).

32. Gershenfeld, 35.

33. Similarly, Ted Sargent views the confluence of chemistry, physics, biology, engineering, and other fields in nanotechnology as a "second renaissance" in which the divergent branches of the tree of knowledge are reconverging. Ted Sargent, *The Dance of Molecules: How Nanotechnology Is Changing Our Lives* (New York: Thunder's Mouth Press, 2006), xiii–xiv.

34. Gershenfeld, 8.

Index

Abstract ideas, patent law and, 131–33, 151, 165

Abstraction: in computer science, 38–41, 231*n*3; patent law and, 127–33, 147–48; problem solving and, 177–79

Abstract wishes, 55

Ada programming language, 36

Affinnova, 196

Agile programming, 70

Agouron Pharmaceuticals, 60

Aiken, Howard, 22, 35

Amazon.com, 110

American Inventors Protection Act, 252*n*15

America Online, 139

Analytical Engine, 36, 49, 230*n*7, 231*n*21, 231*n*22

Antenna design, 1, *2*, 59, 80, 89, 226*n*7, 238*n*36

Apple, 249*n*25

Applications: defined, 138; innovation and, 142–43; patent law and, 144–55; platforms vs., 146

Applied science, 165, 168–71

Art. *See* Field of technology (art); Prior art

Artificial intelligence, 80, 83, 228*n*28, 238*n*47. *See also* Computers: mind mimicked by

Artificial inventions, writing patents for, 201–2

Artificial invention technology: benefits of, 3–4; business and people skills valuable for, 181–84; characteristics

of, 16–18; contemporary, 51–61; cost savings from, 192–94; defined, 225*n*2; examples of, 1; function of, 2–3; future of, 186–90, 211–17; human inventors and, 3–5, 80, 86–87, 90, 177–90, 194, 226*n*9, 241*n*20; innovation and, 144; and invention-description gap, 159–60; invention of, 179–80; job and employment effects of, 180, 184–87; laypersons' use of, 211–17; open business model and, 194–97; outsourcing and, 194–95; patent law and, 8–11, 101–12, 135–36, 148, 151–55, 157–74, 201–10; qualitative difference of, 2; resistance to, 199–200; timing of entry into, 197–99. *See also* Automated inventing; Genies

Artificial selection, 55–61, 181, 200

Art units, 164–65

Assemblers, 42

Assembly, 42

Assembly languages, 40–43

Association for Computing Machinery (ACM), 48

Astronomers, 48–49

AT&T, 168–69

Augmentation, of inventive powers, 88–90, 177–79

Augustin, Larry, 248*n*10

Autodesk, 46, 125, 237*n*21

Autoflow, 91